THW MIND ART

The Mind Art

Lahari Korutla

Published by Lahari Korutla, 2024.

While every precaution has been taken in the preparation of this book, the publisher assumes no responsibility for errors or omissions, or for damages resulting from the use of the information contained herein.

THE MIND ART

First edition. December 15, 2024.

Copyright © 2024 Lahari Korutla.

Written by Lahari Korutla.

Table of Contents

Introduction 1

Chapter 1: Understanding the Mindset Phenomenon 3

Chapter 2: The Power of Belief 8

Chapter 3: Embracing the Growth Mindset 12

Chapter 4: The Art of Positive Thinking 15

Chapter 5: The Transformative Power of Gratitude 18

Chapter 6: Emotional Intelligence as a Key to Success 20

Chapter 7: Mindfulness and Focus in a Distracted World 23

Chapter 8: Resilience: Bouncing Back from Adversity 28

Chapter 9: Creating Lasting Change Through Action 32

Chapter 10: The Role of Habit Formation in Change 36

Chapter 11: The Journey of Lifelong Learning 40

Chapter 12: The Science of Motivation 46

Chapter 13: The Art of Communication 50

Chapter 14: The Influence of Self-Talk 57

Chapter 15: Finding Your Purpose 62

Chapter 16: The Journey of Self-Discovery 67

Chapter 17: Celebrating Progress and Success 73

Conclusion: 77

Introduction

The journey into the world of mindsets begins with a fundamental question: Why do some people navigate life's challenges with ease, while others seem to struggle? The answer lies not in luck or innate talent but in the transformative power of mindset.

The Mind Art invites you to delve into the depths of human cognition and explore its profound impact on every aspect of our lives. In a world brimming with distractions and mounting pressures, mastering the art of harnessing the mind's power has never been more essential.

This book is your roadmap—a guide to unlocking untapped potential, reshaping beliefs, and creating the life you aspire to live. Through insights and practical wisdom, *The Mind Art* empowers you to step into a world where the mind becomes your greatest ally.

As an author deeply rooted in psychology and neuroscience, the exploration reveals the transformative nature of mindset mastery. Countless stories of individuals who have reshaped their mental frameworks illuminate the text. From understanding the interplay of thoughts and beliefs to practical applications, the book distills complex concepts into an actionable guide for personal and professional growth.

What sets this book apart is its holistic approach. It interweaves diverse aspects of mental well-being, encompassing belief systems, emotional intelligence, and mindfulness. The narrative emphasizes that developing a powerful mindset involves more than mere positive thinking; it encompasses a thorough transformation of cognitive habits. As Albert Einstein once said, "The measure of intelligence is the ability to change." By understanding how beliefs shape perceptions and experiences, individuals can take control of their mental landscape.

Central themes emerge, beginning with the concept of a growth mindset. This idea posits that intelligence and abilities are not fixed traits but can be cultivated through effort and perseverance. Embracing challenges as opportunities for growth fosters resilience and enables individuals to thrive in various circumstances. As the book unfolds, it becomes clear that mindset is not a static attribute but a dynamic force that can be refined over time.

LAHARI KORUTLA

This introduction serves as a call to action, encouraging a shift in perspective. Like a master artist creating a masterpiece, each individual has the power to shape their mental canvas. This transformative journey toward a more fulfilling and purposeful life begins with understanding the art of the mind.

Chapter 1: Understanding the Mindset Phenomenon

The exploration of the mindset phenomenon reveals a transformative principle that governs how individuals perceive, interpret, and respond to life's circumstances. Mindset serves as a mental lens through which we filter our experiences, and its profound influence determines not only how we navigate challenges but also how we define success. By delving into the dichotomy between fixed and growth mindsets, this chapter lays the foundation for understanding the critical role mindset plays in shaping life's outcomes.

At the heart of the concept is the distinction between a fixed mindset and a growth mindset, terms popularized by psychologist Carol Dweck. A fixed mindset assumes that abilities, intelligence, and talents are innate and unchangeable. Those with a fixed mindset often fear failure, as it challenges their sense of self-worth. In contrast, a growth mindset views intelligence and abilities as malleable, capable of expansion through effort, persistence, and learning. This perspective not only encourages individuals to embrace challenges but also fosters resilience in the face of adversity.

The power of mindset is brought to life through the story of two siblings with contrasting perspectives. The older sibling, burdened by a fixed mindset, avoids risks and views failure as a definitive sign of inadequacy. Meanwhile, the younger sibling, driven by a growth mindset, sees challenges as puzzles to solve and opportunities for learning. When faced with a failed science project, the older sibling abandons the task, believing they lack the skill, while the younger sibling seeks guidance, learns from the failure, and eventually excels. This anecdote underscores the pivotal role mindset plays in determining outcomes.

A vivid historical example further illuminates this concept: the story of Thomas Edison and his relentless pursuit of invention. Despite thousands of failed attempts to create the light bulb, Edison famously remarked, "I have not failed. I've just found 10,000 ways that won't work." His perspective exemplifies the growth mindset, turning what could have been insurmountable failures into stepping stones to groundbreaking success. Edison's unwavering belief in progress, fueled by a mindset oriented toward learning and perseverance, demonstrates that failure is not a dead-end but a critical component of the journey to achievement.

To apply these principles in everyday life, exercises and reflective practices are essential. One activity involves examining personal experiences with failure and identifying moments where a growth mindset could have reframed the outcome. For example, consider a student struggling with public speaking. A fixed mindset might label them as "not a natural speaker," discouraging them from trying. However, by adopting a growth mindset, the student could view each presentation as practice, steadily improving their confidence and skills over time.

Let's consider another real-world scenario: a professional encountering repeated setbacks in launching a new project. With a fixed mindset, the individual might view these failures as evidence of incompetence, ultimately abandoning their goals. However, a growth mindset encourages them to analyze the setbacks, learn from their mistakes, and refine their strategies. As they persist, they develop resilience and adaptability—key traits that pave the way for eventual success.

This chapter also highlights how fixed mindsets develop and how they can be challenged. Often, these limiting beliefs are rooted in early experiences, societal expectations, or the fear of judgment. For example, a child repeatedly told they are "not good at math" might internalize this label, avoiding the subject and solidifying a fixed mindset. Overcoming this requires intentional effort to reframe these beliefs. Cognitive flexibility exercises, such as practicing reframing negative self-talk into positive affirmations, help individuals shift their perspective. For instance, replacing "I can't do this" with "I can learn to do this" rewires thought patterns and fosters a more adaptive mindset.

Mindsets are not set in stone. Research in neuroscience supports the idea of neuroplasticity, which is the brain's ability to reorganize itself by forming new neural connections. This means individuals can transition from a fixed mindset to a growth-oriented one through deliberate practice and effort. As one begins to challenge limiting beliefs and embrace discomfort as a natural part of growth, they dismantle self-imposed barriers and unlock their potential.

Another compelling example is drawn from the realm of sports. Consider the journey of Michael Jordan, widely regarded as one of the greatest basketball players of all time. Early in his career, Jordan was cut from his high school basketball team. A fixed mindset might have led him to believe he wasn't cut out for the sport, but his growth-oriented perspective pushed him to work harder, refine his skills, and ultimately achieve unparalleled success. His story is a testament to how embracing challenges and persevering through failure can lead to greatness.

The chapter also explores how societal and cultural factors can reinforce or hinder mindset development. Educational systems, for instance, often prioritize performance over effort, inadvertently fostering fixed mindsets. A student praised solely for their natural intelligence may come to fear challenges that could expose perceived weaknesses. To counter this, educators and parents can emphasize the value of effort and persistence, cultivating a growth-oriented environment that encourages lifelong learning.

As the chapter concludes, it underscores the transformative power of mindset in reshaping lives. By embracing a growth mindset, individuals can unlock their potential, develop resilience, and view challenges as opportunities for growth. The mindset phenomenon, once understood and applied, becomes a compass guiding us toward fulfillment and success. It is not just about achieving specific goals but about adopting a perspective that enriches every facet of life.

"Challenges are what make life interesting; overcoming them is what makes life meaningful." – Joshua J. Marine

LAHARI KORUTLA

Through deliberate effort, reflection, and a commitment to growth, the art of mastering one's mindset becomes a lifelong journey—a journey that begins with understanding and embracing the mindset phenomenon.

THE MIND ART

LAHARI KORUTLA
Chapter 2: The Power of Belief

Beliefs hold a remarkable power to shape perceptions, guide behavior, and ultimately determine the outcomes of our lives. From the way we view ourselves to the goals we pursue, the foundation of belief acts as a compass, directing us toward success or limiting our potential. The chapter begins with the inspiring story of an athlete who defied physical limitations through sheer belief in their abilities. Despite being told they would never run again after a severe injury, the athlete visualized recovery and embraced the belief that they could heal. Over time, they not only regained mobility but went on to win multiple marathons. This example highlights how belief serves as either the greatest barrier or the most powerful catalyst to achievement.

Beliefs are deeply ingrained, often formed during childhood when we are most impressionable. They stem from familial teachings, cultural norms, societal expectations, and early experiences. For instance, a child repeatedly told they lack talent in a particular area may internalize this belief and carry it into adulthood, avoiding opportunities for growth in that domain. These limiting beliefs act as invisible chains, constraining potential and reinforcing negative cycles. A reflective exercise included in this chapter encourages individuals to identify these deeply rooted beliefs, asking them to examine their origins and question whether they serve or obstruct personal progress.

The concept of self-fulfilling prophecies illustrates how belief systems manifest into reality. When we believe something about ourselves—whether positive or negative—we tend to act in ways that validate that belief. The story of a struggling musician exemplifies this principle. Initially plagued by self-doubt, they believed their lack of natural talent would prevent success. However, a shift in mindset, sparked by the realization that effort could surpass innate ability, led them to practice relentlessly. Over time, their confidence grew, their skill improved, and they went on to achieve significant acclaim. This transformation reinforces Mahatma Gandhi's wisdom:

> "Your beliefs become your thoughts, your thoughts become your actions, your actions become your habits, your habits become your values, and your values become your destiny."

The chain reaction described in Gandhi's quote underscores the importance of cultivating empowering beliefs. When belief systems are supportive, they set off a cascade of positive thoughts, behaviors, and outcomes.

THE MIND ART

Practical strategies for reshaping beliefs emerge as crucial tools for personal transformation. Visualization, for example, involves imagining oneself successfully achieving a goal. Athletes often use this technique, mentally rehearsing winning a race or scoring a goal, which primes their minds for success. Similarly, positive affirmations—statements like "I am capable of achieving greatness" or "I grow stronger with every challenge"—help replace disempowering thoughts with constructive ones. These strategies train the brain to reframe limitations, turning perceived obstacles into opportunities for growth.

Consider the story of Oprah Winfrey, a global icon who overcame a difficult childhood marked by poverty and abuse. She often speaks of how the belief in her own worth, despite external circumstances, propelled her forward. Her story serves as a testament to the transformative power of belief systems. By choosing to focus on her potential rather than her challenges, she rewrote her narrative and achieved extraordinary success.

Belief systems are not static; they are malleable and can evolve with conscious effort. Case studies of prominent figures who broke free from limiting beliefs further illustrate this point. For instance, Albert Einstein, often misunderstood as a child and deemed "slow" by teachers, rejected these limiting perceptions. Instead, he cultivated a belief in his intellectual curiosity and potential, which ultimately led to groundbreaking contributions to science. These examples serve as a reminder that beliefs, when aligned with our goals, can be powerful tools for transformation.

To solidify the understanding of belief 's influence, the chapter introduces guided exercises designed to challenge limiting beliefs. One such exercise involves listing current beliefs about oneself and categorizing them as empowering or limiting. For each limiting belief, individuals are encouraged to write a counter-belief that reframes the original thought in a positive light. For example, replacing "I'm not good at public speaking" with "I can improve my public speaking skills with practice." This shift fosters a mindset of growth and possibility.

The chapter also explores how societal expectations can either reinforce or dismantle personal beliefs. For example, a teacher's encouragement can instill confidence in a hesitant student, just as negative stereotypes can perpetuate self-doubt. Recognizing these external influences allows individuals to reclaim agency over their belief systems, choosing which messages to internalize and which to discard.

Ultimately, the chapter underscores that beliefs are not immutable; they can be challenged, reshaped, and aligned with personal aspirations. By consciously choosing empowering beliefs, individuals unlock a mindset rooted in resilience, optimism, and growth. Just as the athlete, musician, and countless others have demonstrated, the power of belief is a transformative force that has the potential to rewrite the trajectory of any life.

"Man is what he believes." – Anton Chekhov

Belief, as this chapter concludes, is not merely a passive construct but an active tool. By understanding its power and consciously shaping it, we can create lives that reflect our highest potential. The journey to mastering belief begins with awareness, progresses through intentional practice, and culminates in a mindset that supports boundless growth and achievement.

THE MIND ART

LAHARI KORUTLA
Chapter 3: Embracing the Growth Mindset

A growth mindset is a powerful lens through which one can view life's challenges as opportunities for improvement and learning. This chapter opens with the story of a renowned artist who faced countless rejections early in their career. Despite the repeated "no's," they refused to equate rejection with failure. Instead, they interpreted each critique as a stepping stone toward mastery. By embracing feedback and honing their craft, the artist eventually gained global recognition. This narrative serves as a vivid reminder that true growth often happens beyond the boundaries of comfort zones.

Central to the growth mindset is the idea that effort, rather than innate ability, is the cornerstone of achievement. Consider the story of a student who initially struggled in mathematics. Labeled as "average," they were tempted to accept this limitation. However, inspired by a teacher who encouraged them to see mistakes as part of the learning process, the student began dedicating extra time to understanding the subject. Slowly but steadily, their confidence grew. By the end of the academic year, they not only excelled in math but developed a love for problem-solving. This transformation highlights how perseverance and the willingness to embrace effort can lead to profound changes.

To deepen the understanding of the growth mindset, reflective exercises are introduced. Readers are encouraged to identify areas in their lives where they feel stuck or discouraged. By reframing these obstacles as opportunities for growth, they can begin to shift their perspective. For example, someone struggling with public speaking might see each presentation not as a test of ability but as a chance to refine their skills. This approach fosters resilience, turning setbacks into valuable learning experiences.

A young entrepreneur, after launching their first startup, encountered several failures—products that didn't sell, partnerships that dissolved, and financial losses. Initially, they saw these failures as proof of inadequacy. However, after learning about the growth mindset, they chose to analyze each setback, asking what could be improved or done differently. With each subsequent venture, they applied these lessons, eventually building a thriving company. Their story illustrates that a growth mindset not only helps navigate adversity but also equips individuals with the tools to learn and adapt.

Cognitive flexibility, or the ability to shift perspectives, is a cornerstone of the growth mindset. Albert Bandura's assertion that "People's beliefs about their capabilities have a major effect on their success" underscores the importance of this adaptability. Practical exercises in this chapter guide readers to practice seeing situations from multiple viewpoints. For instance, a professional facing criticism at work might initially feel disheartened. However, by viewing the feedback as an opportunity to enhance their skills, they can transform a potentially negative experience into a constructive one. This adaptability is crucial in both personal and professional spheres.

A significant component of the growth mindset is setting process-oriented goals rather than focusing solely on outcomes. When individuals shift their attention to the effort and learning involved in achieving a goal, they reduce the fear of failure and cultivate a sense of accomplishment with every small step. For example, a runner training for a marathon might celebrate completing each weekly run rather than fixating on the race day itself. This incremental approach not only sustains motivation but also fosters a deeper appreciation for the journey of growth.

Consider the life of J.K. Rowling, the famed author of the *Harry Potter* series. Before achieving literacy success, she faced numerous rejections from publishers, financial struggles, and personal challenges. Instead of succumbing to despair, she persisted, refining her manuscript and believing in her story. Her journey exemplifies the growth mindset: resilience, continuous improvement, and a focus on the process rather than immediate outcomes.

The chapter also emphasizes the importance of celebrating progress, no matter how small. This celebration reinforces the idea that growth is a journey, not a destination. For instance, learning a new skill like painting or playing an instrument might initially feel daunting. However, by acknowledging each incremental improvement—whether it's mastering a basic stroke or playing a simple melody—individuals maintain their enthusiasm and drive.

In conclusion, cultivating a growth mindset is an ongoing journey that demands self-awareness, effort, and adaptability. By embracing challenges and viewing them as opportunities to learn, individuals unlock new levels of resilience and satisfaction. The transformative power of a growth mindset lies not only in achieving specific goals but also in fostering a lifelong love for learning and self-improvement. This mindset lays the groundwork for navigating future challenges with confidence and optimism, ensuring that every experience becomes a stepping stone toward personal and professional fulfillment.

As this chapter concludes, it encourages readers to internalize the words of Carol Dweck, the psychologist who popularized the concept of the growth mindset:

"No matter what your ability is, effort is what ignites that ability and turns it into accomplishment."

By embracing this belief, individuals can begin to view their potential as limitless, with each step forward no matter how small leading to meaningful growth and transformation.

Chapter 4: The Art of Positive Thinking

Positive thinking is more than a feel-good concept; it is a powerful approach that influences how individuals perceive and respond to life's challenges. This chapter opens with a narrative of a single mother who faced numerous hardships, including financial struggles and health issues. Rather than succumbing to despair, she focused on small blessings—a kind word from a stranger, her child's laughter, or even the simple beauty of a sunset. This shift in perspective didn't erase her difficulties but gave her the strength to navigate them with resilience and hope. Her story illustrates the core principle of positive thinking: it's not about avoiding reality but choosing how to respond to it.

The psychology of positive thinking reveals its profound impact on mental health and overall life satisfaction. Scientific studies underscore that individuals who maintain an optimistic outlook are more likely to experience lower stress levels, improved cardiovascular health, and even a longer lifespan. For example, a landmark study conducted at Harvard University found that individuals with a positive mindset were 31% more productive and had 23% fewer health-related complications. These findings validate that positive thinking isn't merely about "looking on the bright side" but is deeply rooted in tangible benefits for well-being and success.

Norman Vincent Peale's timeless wisdom, "Change your thoughts and you change your world," encapsulates the transformative power of mindset. When individuals consciously choose positive thoughts, they influence not only their inner world but also how they interact with the external environment. For instance, a job seeker who views rejection as a stepping stone rather than a failure is more likely to persist, refine their approach, and ultimately achieve success. This subtle yet profound shift in perspective changes the trajectory of experiences and outcomes.

The Power of Gratitude

Consider the example of a young executive who was overwhelmed by the pressures of her career. She constantly focused on what went wrong—missed deadlines, critical feedback, or lost opportunities. Her stress affected her productivity and personal life. One day, she began practicing gratitude journaling. Each evening, she wrote down three things that went well during the day, no matter how small. Over time, she noticed a significant change in her mindset. Instead of fixating on the negatives, she began to appreciate her achievements, support from her team, and moments of joy. This shift not only reduced her stress but also improved her relationships and performance at work.

To cultivate positive thinking, the chapter introduces practical techniques that can be seamlessly integrated into daily life. Gratitude journaling, as highlighted above, is one of the most effective methods. Writing down positive experiences trains the brain to focus on the good, fostering an attitude of appreciation. Another powerful tool is daily affirmations—simple yet impactful statements like "I am capable," "I attract positivity," or "I can overcome challenges." Repeating affirmations helps rewire the brain, reinforcing self-belief and optimism.

Real-life success stories add depth to the discussion, illustrating how positivity fuels actionable change. One such story involves a man who lost his job during an economic downturn. Initially devastated, he chose to shift his focus from loss to opportunity. He took the time to upskill, network, and eventually found a more fulfilling role in a different industry. His optimism not only helped him persevere but also inspired those around him to adopt a similar mindset.

The role of social environments in shaping positivity is another crucial aspect explored in this chapter. Surrounding oneself with supportive and optimistic individuals creates a feedback loop of encouragement and growth. Conversely, toxic environments can drain emotional energy and reinforce negativity. Strategies for cultivating a positive social circle include spending time with uplifting people, seeking out mentors, and participating in communities that align with personal values. For instance, joining a group focused on volunteer work can foster a sense of purpose and connection while promoting positivity.

The Science of Positive Influence

Research shows that positivity is contagious. When individuals radiate optimism, they influence those around them, creating a ripple effect. A study conducted by the University of California demonstrated that individuals in happier social networks were 25% more likely to experience improved emotional well-being themselves. This insight emphasizes the dual role of positivity—not only enhancing one's life but also uplifting the broader community.

As the chapter progresses, it underscores that positive thinking is a skill, not an inherent trait. Like any skill, it requires practice, intention, and patience. Readers are encouraged

THE MIND ART

to start small, focusing on daily wins and gradually expanding their capacity to see the good in every situation. It's a journey of rewiring mental habits, much like physical exercise strengthens the body.

In conclusion, positive thinking emerges as a transformative art that empowers individuals to face life's adversities with grace and resilience. By cultivating gratitude, practicing affirmations, and surrounding oneself with positivity, individuals can significantly enhance their emotional well-being and life satisfaction. Moreover, as optimism spreads from one person to another, it creates a ripple effect, inspiring communities to adopt a brighter, more hopeful perspective.

As Mahatma Gandhi wisely stated, "Keep your thoughts positive because your thoughts become your words. Keep your words positive because your words become your behavior. Keep your behavior positive because your behavior becomes your habits. Keep your habits positive because your habits become your values. Keep your values positive because your values become your destiny." This timeless wisdom encapsulates the essence of the journey toward positive thinking, reminding us that our mindset shapes our destiny.

As this chapter closes, it leaves readers with a powerful reminder:

"You cannot control what happens to you, but you can control how
you respond to it. And in that response lies your power."

By embracing the art of positive thinking, individuals not only transform their lives but also contribute to a world filled with hope, strength, and possibility.

LAHARI KORUTLA
Chapter 5: The Transformative Power of Gratitude

Gratitude is a cornerstone of emotional resilience and mental well-being, serving as a bridge between challenges and fulfillment. This chapter begins with the moving story of a war veteran who found solace and strength by focusing on daily gratitude during his recovery from physical injuries and emotional trauma. His journey underscores the transformative nature of gratitude, shifting attention from loss to abundance and fostering hope in the midst of despair. As Dr. Robert Emmons eloquently stated, "Gratitude is not only the greatest of virtues but the parent of all the others."

The Science of Gratitude

Gratitude is not just an abstract virtue; it is rooted in science. Research by positive psychology experts demonstrates that gratitude practices can significantly enhance mental health. Studies show that individuals who regularly practice gratitude report better sleep, improved physical health, reduced stress levels, and a heightened sense of purpose. These findings highlight that gratitude has the potential to rewire the brain, helping individuals cultivate a more positive outlook on life.

To illustrate this, consider the case of a young entrepreneur struggling with financial losses. Initially overwhelmed by failure, she began a daily gratitude journal, noting three things she appreciated each day. Over time, this practice shifted her focus from what she lacked to the resources, relationships, and opportunities still available to her. This subtle but powerful transformation helped her rebuild her business with a renewed sense of optimism and resilience.

Practical Gratitude Practices

The chapter offers a variety of strategies to weave gratitude into everyday life.

- Gratitude Journals: Writing down three things you are grateful for each day fosters reflection and appreciation for the present moment.

- Expressing Gratitude: Whether through handwritten notes, spoken words, or simple acts of kindness, expressing gratitude strengthens bonds and fosters deeper connections.

THE MIND ART

- Mindful Gratitude Exercises: Practicing mindfulness to focus on what's going well in the moment enhances emotional well-being.

A relatable example is introduced here: a working parent juggling multiple responsibilities finds peace in a nightly gratitude ritual with their children. Each family member shares one thing they're thankful for, creating a positive end to their day and strengthening family bonds. Gratitude Challenges

The idea of structured gratitude challenges is also explored as a way to make gratitude a consistent habit. For instance, a "30-Day Gratitude Challenge" encourages participants to write down or verbally acknowledge one thing they are grateful for each day. The cumulative effect of such a practice fosters a shift in mindset, making gratitude an integral part of life.

The Ripple Effect of Gratitude

Gratitude doesn't only benefit the individual; its effects ripple outward, impacting communities and relationships. A school teacher, for example, implemented gratitude activities in her classroom, encouraging students to appreciate one another's contributions. This practice not only improved the students' morale but also created a more supportive and empathetic learning environment.

In closing, gratitude is not merely a reaction to favorable circumstances; it is a deliberate choice that transforms lives. As the ancient philosopher Cicero asserted, "Gratitude is not only the greatest virtue but the parent of all others." By choosing gratitude, individuals invite joy, resilience, and deeper connections into their lives.

Gratitude becomes the lens through which one views the world, magnifying blessings and minimizing struggles. It is a daily practice that reshapes perspectives, reinforces values, and nurtures the soul. By embracing the transformative power of gratitude, individuals unlock profound fulfillment, no matter their circumstances.

"When we focus on our gratitude, the tide of disappointment goes out, and the tide of love rushes in." – Kristin Armstrong

Chapter 6: Emotional Intelligence as a Key to Success

Emotional intelligence (EI) is increasingly recognized as a cornerstone of success, shaping the way individuals navigate personal and professional relationships. The chapter opens with the story of a corporate leader who turned around a struggling team not by focusing solely on numbers but by fostering an environment of trust, understanding, and emotional connection. This narrative demonstrates how EI equips individuals to navigate challenges with empathy and adaptability, enabling them to inspire, lead, and collaborate effectively.

The Core Components of Emotional Intelligence

Breaking down the framework of EI, the chapter delves into its core components: self-awareness, self-regulation, motivation, empathy, and social skills.

- Self-awareness serves as the foundation, enabling individuals to understand their emotional triggers and reactions.

- Self-regulation builds upon this awareness, emphasizing the ability to manage impulses and adapt to changing circumstances without becoming overwhelmed.

- Motivation reflects the drive to achieve goals, even in the face of setbacks.

- Empathy is the ability to see the world through another's eyes, creating bonds of understanding.

- Social skills tie everything together, allowing individuals to communicate effectively, resolve conflicts, and build lasting relationships.

These components collectively shape a person's ability to navigate emotional landscapes, both internally and in interactions with others. Daniel Goleman's assertion, "In a very real sense, we have two minds, one that thinks and one that feels," emphasizes the importance of harmonizing intellect with emotion in decision-making.

The Importance of Emotional Intelligence in Leadership

Leadership provides a powerful lens through which the significance of EI can be understood. A compelling example comes from the story of a school principal who turned a failing institution into a thriving community. By actively listening to teachers, empathizing with parents, and encouraging students, the principal demonstrated how emotionally intelligent leadership fosters trust and collaboration. This transformation underscores the idea that EI is not just about understanding one's own emotions but also about recognizing and influencing the emotional climate of others.

Practical Steps to Develop Emotional Intelligence

The chapter introduces several strategies for enhancing EI:

1. Journaling: Encourages self-reflection and awareness by documenting emotional responses to daily events.
2. Mindfulness practices: Help individuals stay present and regulate their emotions effectively.
3. Role-playing scenarios: Allow individuals to practice empathy by putting themselves in others' shoes.

For instance, a reflective exercise guides individuals to recall a recent conflict and analyze it through the lens of EI. Questions such as "What emotions were involved?" and "How could the outcome have been improved through empathy or self-regulation?" offer actionable insights into emotional dynamics.

Empathy as the Heart of Emotional Intelligence

Empathy is presented as the heart of EI, bridging the gap between self-awareness and social skills. Brené Brown's poignant statement, "Empathy is seeing with the eyes of another, listening with the ears of another, and feeling with the heart of another," serves as a guiding principle. Practical techniques, such as active listening and open-ended questioning, are discussed as ways to deepen empathetic connections.

A real-life example illustrates this: A customer service representative, initially overwhelmed by an irate client, shifts the conversation by validating the client's frustration and offering genuine support. The outcome transforms a hostile exchange into a collaborative resolution, highlighting the transformative potential of empathy in real-world situations.

The Role of Emotional Intelligence in Team Dynamics

The narrative further explores how EI enhances teamwork. High-EI individuals contribute to healthier team dynamics by mediating conflicts, fostering inclusivity, and promoting open communication. Case studies of successful teams reveal that emotional intelligence is often the distinguishing factor between teams that merely function and those that excel.

Conclusion

The chapter concludes by reiterating that emotional intelligence is a skill, not an inherent trait, and can be cultivated through intentional practice. By developing EI, individuals gain tools to navigate life's complexities with grace and effectiveness, forging deeper connections and achieving greater success. Emotional intelligence, as the chapter demonstrates, is not merely an asset; it is a transformative force that empowers individuals to thrive in an interconnected world

Chapter 7: Mindfulness and Focus in a Distracted World

In an era defined by rapid technological advancements and constant connectivity, the practice of mindfulness has become more important than ever. With the pervasive distractions of social media, emails, and the general hustle and bustle of modern life, it's easy to feel disconnected, overwhelmed, and stretched thin. Yet, amid this chaos, mindfulness emerges as a powerful tool for reclaiming focus and clarity, offering individuals the chance to reestablish a sense of presence and balance.

The Story of a Mindful Journey

Consider the story of Emily, a busy professional, who had grown accustomed to the constant barrage of notifications and demands. Her days were a blur of meetings, tasks, and responsibilities. Despite her achievements, Emily often found herself feeling drained, anxious, and disconnected. It wasn't until she began exploring mindfulness that her perspective began to shift.

Emily started with small steps—taking five minutes each morning to sit quietly and breathe deeply, focusing only on the sensation of her breath entering and leaving her body. She noticed that, after a few weeks, her mornings became more intentional and calm. As her practice grew, she began to incorporate mindfulness into her daily activities. Whether walking to the office, answering emails, or even talking to colleagues, Emily made a conscious effort to stay present in each moment. What she discovered was not just a reduction in stress, but a deeper sense of connection to herself and the world around her.

> This narrative highlights a powerful truth: mindfulness doesn't require dramatic shifts in our lives. Sometimes, all it takes is a small commitment to being more present, one moment at a time.

The Science Behind Mindfulness

The benefits of mindfulness are backed by substantial scientific research. Studies have shown that mindfulness practices, such as meditation and mindful breathing, can significantly improve mental health by reducing symptoms of anxiety, depression, and stress. Mindfulness has also been shown to improve emotional regulation, allowing individuals to respond to challenges with greater patience and clarity, rather than reacting impulsively or emotionally.

A key concept in mindfulness is its ability to reduce cognitive overload. In today's fast-paced world, our brains are constantly bombarded with information. This flood of stimuli can lead to mental exhaustion, hinder focus, and impair decision-making. By training ourselves to be mindful, we create the mental space needed to process and prioritize information. As Jon Kabat-Zinn, one of the pioneers of mindfulness research, defines it, "Mindfulness is the awareness that arises from paying attention, on purpose, in the present moment, and non-judgmentally." This definition encapsulates mindfulness as a way to ground ourselves in the present, where we can choose how to respond to life's challenges, instead of being swept away by them.

Practical Mindfulness Techniques

Mindfulness is not a one-size-fits-all solution. Different techniques can be used depending on the individual's needs and preferences. Some of the most effective practices include:

1. Mindful Breathing:

One of the simplest and most accessible mindfulness techniques is mindful breathing. The practice involves focusing on the breath, observing the inhale and exhale, without trying to control it. This practice can be done anywhere—while waiting for the bus, during a break at work, or before a stressful meeting. A simple mindful breathing session of five to ten minutes can bring a profound sense of calm and clarity.

Daniel, a father of two, found himself increasingly stressed by the demands of work and family life. After learning about mindful breathing, he decided to dedicate ten minutes each morning before his children woke up to sit quietly and focus on his breath. This practice became a sanctuary for him—a moment of peace before diving into the chaos of the day. Over time, Daniel noticed a greater sense of patience with his family and more focus at work.

2. Body Scan Meditation:

Body scan meditation is a technique where individuals mentally scan their bodies, from head to toe, noting any sensations, tension, or discomfort. This practice helps individuals reconnect with their bodies and become more aware of where they are holding stress. It is especially effective for people who experience chronic stress or anxiety, as it helps identify and release physical tension.

3. Mindful Walking:

Mindful walking is a simple practice where an individual walks slowly and deliberately, focusing on each step and the sensations of movement. It can be done in nature, in a park, or even in the office hallways. This practice can help individuals who find it difficult to sit still, allowing them to ground themselves in the present moment through physical activity.

Anna, a writer struggling with deadlines, began taking daily mindful walks around the block during her breaks. She found that the practice not only provided a mental reset but also helped her overcome writer's block. By focusing solely on the act of walking, her mind became clearer, and her creativity flourished.

4. Mindful Observation:

Mindful observation encourages individuals to focus on their surroundings, paying attention to details they may usually overlook. Whether it's watching the play of light on a surface or observing the movement of leaves in the wind, this practice cultivates a deeper appreciation for the present moment.

After incorporating mindful observation into her daily routine, Maya, an overworked lawyer, found that she became more attuned to the world around her. Rather than rushing through her lunch breaks, she started sitting outside, paying attention to the birds, the wind, and the clouds. These small moments of connection with her environment helped her feel less overwhelmed and more at peace.

Mindfulness in Action

The true beauty of mindfulness lies in its adaptability. It can be practiced anywhere and at any time. Whether you're at work, at home, or on the go, the key is to take a moment and reconnect with the present. In the workplace, mindfulness can improve focus, reduce stress, and enhance communication. It allows individuals to approach tasks with greater clarity, creativity, and efficiency. For example, a manager who practices mindfulness may approach a conflict with a colleague with greater empathy, understanding that a calm, measured response will be more effective than reacting impulsively.

In relationships, mindfulness fosters better communication and deeper connections. By being present and listening attentively, we show others that we value their thoughts and feelings. This creates a foundation of trust and mutual respect, which strengthens personal and professional relationships.

Real-Life Success Stories

Across various fields, many individuals have credited mindfulness with helping them overcome challenges and improve their quality of life. Take the story of Sarah, a teacher who struggled with burnout. She incorporated mindfulness practices into her daily routine, taking a few minutes during breaks to reset. Over time, Sarah noticed an improvement in her ability to manage stress and a deeper connection to her students. Her relationships with her colleagues also became more harmonious, as she was able to approach difficult conversations with empathy and understanding.

Similarly, athletes like Novak Djokovic have publicly shared their mindfulness practices as part of their mental preparation. Djokovic uses mindfulness and meditation techniques to enhance his focus and mental clarity, which have contributed to his success on the tennis court.

The Benefits of Mindfulness for Focus

Mindfulness is particularly effective in enhancing focus. In a world filled with constant distractions, the ability to direct attention and sustain concentration is a valuable skill. Research has shown that mindfulness training improves cognitive abilities, such as attention span, memory, and the ability to switch between tasks efficiently. With regular practice, individuals can develop a stronger capacity for sustained focus, leading to higher productivity and improved performance.

Conclusion: A Path to Greater Clarity and Focus

In conclusion, mindfulness is a transformative practice that offers individuals the tools to navigate life's distractions and stressors with grace and clarity. By cultivating mindfulness, individuals can enhance their focus, reduce anxiety, and foster deeper connections with themselves and others. Whether through mindful breathing, body scan meditation, or simply observing the world around us, mindfulness has the power to bring greater peace and purpose to our daily lives. As we continue to face a world full of distractions, mindfulness serves as an antidote, allowing us to slow down, be present, and find balance in the midst of it all.

Chapter 8: Resilience: Bouncing Back from Adversity

Resilience is often described as the ability to bounce back from adversity, to rise above setbacks, and to emerge stronger than before. In a world that is ever-changing and full of challenges, resilience is an essential trait that allows individuals to persevere and grow, regardless of what obstacles they face. While some may view resilience as an inherent quality, it is, in fact, a skill that can be developed over time with intentional effort and the right mindset.

The Power of Resilience: A Story of Triumph

Consider the story of David, a man who experienced a series of life-altering events. At the age of 30, David was suddenly diagnosed with a rare illness that threatened his mobility and independence. This was just the beginning of a period filled with personal loss, financial struggles, and health setbacks. Despite the overwhelming challenges, David found a way to adapt. Through sheer determination and the development of new coping strategies, he not only learned to manage his illness but also started advocating for others facing similar challenges.

David's journey highlights the essence of resilience: the capacity to move forward in the face of adversity. While it was not an easy path, it was a transformative one. By embracing change, focusing on what he could control, and seeking support, David demonstrated that resilience is about growth through adversity—not just survival. His story serves as a reminder that, no matter the circumstances, resilience is about finding strength in moments of vulnerability and learning from hardship.

Defining Resilience: Not Just a Trait, But a Skill

Resilience is often misunderstood as something that people either have or don't have, but it is more nuanced than that. It's a dynamic skill that evolves over time and is influenced by various factors, including mindset, coping strategies, and social support. While some people may have a natural inclination toward resilience due to their upbringing or personality, anyone can cultivate this ability with effort and practice.

One key characteristic of resilient individuals is their ability to remain optimistic, even in the face of setbacks. Optimism, however, is not the same as ignoring reality. Rather, it is the ability to focus on potential solutions, the belief that things can improve, and the confidence that challenges can be overcome with perseverance and effort.

Another critical trait of resilience is adaptability. Life is full of uncertainty, and resilient people are those who can adjust to changing circumstances. Adaptability allows individuals to shift their mindset, develop new skills, and adjust their strategies when confronted with challenges. It's not about avoiding difficulties but about embracing them with a mindset that looks for opportunities for growth.

Resilience is also deeply tied to problem-solving skills. The ability to find solutions, even in the most difficult situations, is crucial to navigating adversity. Resilient individuals are often resourceful, creative, and open-minded, seeking out different ways to approach challenges. They understand that setbacks are inevitable but that each obstacle presents an opportunity to learn and grow.

Developing Resilience: Practical Strategies

While resilience may seem like an innate quality in some, it can be developed through intentional practice. Developing resilience starts with building a growth mindset—the belief that abilities and intelligence can be developed through hard work, dedication, and learning. This mindset encourages individuals to view challenges not as insurmountable obstacles but as opportunities for growth.

One practical strategy for building resilience is to reflect on past challenges. Guided exercises can help individuals think about previous hardships and identify the coping mechanisms they used to overcome them. This reflective practice not only provides insight into one's strengths but also highlights areas for growth. Some key questions to consider might include:

- What were the challenges I faced?
- How did I respond to those challenges?
- What strategies helped me cope?

- What did I learn from those experiences?

By answering these questions, individuals can begin to identify their personal resilience strategies and build upon them for future challenges.

Example of Reflection in Action:

Sarah, a teacher, faced a tough year when her school was facing budget cuts, and she had to manage increased workload and pressure from parents and administration. Rather than letting the stress overwhelm her, she took time each week to reflect on her responses to the challenges. Through this reflective practice, she realized that when she communicated openly with her colleagues and sought their support, she felt more empowered and less stressed. Sarah's ability to recognize the role of collaboration in her resilience helped her navigate the rest of the year more effectively. She was able to adopt a mindset that embraced teamwork and shared responsibility, which became a source of strength.

The Role of Self-Compassion in Building Resilience

One of the most powerful ways to build resilience is through self-compassion. This involves treating oneself with the same kindness and understanding that one would offer a friend facing hardship. According to Kristin Neff, a leading researcher in self-compassion, "Self-compassion is simply giving the same kindness to ourselves that we would give to others." In moments of struggle, people often tend to be hard on themselves, which can hinder their ability to cope and move forward. However, by practicing self-compassion, individuals can foster emotional resilience.

Self-compassion allows individuals to acknowledge their suffering without judgment, to accept imperfections, and to embrace their humanity. It reminds people that they are not alone in their struggles and that it's okay to ask for help or take time for self-care. This approach builds emotional strength and reduces feelings of isolation and self-doubt.

Example of Self-Compassion in Action:

Tom, an entrepreneur, faced the failure of his first startup, which caused him significant stress and self-criticism. He was tempted to give up, feeling like he was incapable of succeeding. However, after learning about self-compassion, Tom started practicing it by treating himself with the same kindness he would offer a friend. He acknowledged his disappointment but also recognized the lessons he had learned. Instead of berating himself, Tom used the failure as an opportunity to reflect on what went wrong and how he could improve. This practice of self-compassion allowed him to bounce back with a stronger mindset and, eventually, the confidence to launch a second business that found greater success.

The Importance of Supportive Relationships

Another critical factor in resilience is the support of others. Resilient individuals tend to have strong social connections—whether through family, friends, or colleagues—that provide emotional support and encouragement during difficult times. These relationships serve as a safety net, reminding individuals that they are not alone in their struggles. Having a support system helps individuals stay grounded and provides them with the motivation and perspective needed to push through adversity.

In conclusion, resilience is not a fixed trait but a dynamic skill that can be developed and strengthened over time. It involves cultivating optimism, adaptability, and problem-solving abilities, while also practicing self-compassion and seeking support from others. By learning to embrace challenges, reflect on past experiences, and treat ourselves with kindness during difficult times, we can build greater resilience and emerge stronger from adversity. Like any skill, resilience takes practice and patience, but with time, it becomes an invaluable tool for navigating life's inevitable ups and downs.

Through deliberate effort and mindful practice, we can not only cope with challenges but use them as stepping stones for personal growth and fulfillment. In this way, resilience becomes not just a response to adversity but a lifelong practice that empowers individuals to live more meaningful, successful, and connected lives.

Chapter 9: Creating Lasting Change Through Action

The process of personal transformation often feels like a distant dream, especially when one feels stuck in the routines and challenges of everyday life. However, true change requires more than just good intentions. It demands intentional action, a commitment to moving forward, and a willingness to persist in the face of obstacles. This chapter is dedicated to understanding how individuals can create lasting change through purposeful actions, backed by the right mindset and strategies.

The Power of Action: A Story of Change

Let's begin by considering the story of Laura, a woman in her late 30s who had spent most of her adult life feeling unsatisfied and directionless. Despite achieving career success, Laura felt that something was missing. For years, she wanted to pursue her passion for writing, but the demands of her job and family left her feeling overwhelmed. She often dreamed of writing a book but never took steps toward making that dream a reality.

It wasn't until Laura reached a point where she could no longer ignore the nagging feeling of unfulfilled potential that she decided to take action. She began by writing down her goals and dreams, mapping out what truly mattered to her. From there, she broke down her larger goal of writing a book into smaller, manageable steps. She set aside 15 minutes every morning to write, committing to the process regardless of how busy her day might be. Slowly but surely, Laura made progress. Within two years, she not only finished her book but also became a published author.

> Laura's journey illustrates an important point: change begins with intentional action. Without that initial step toward the goal, her dreams would have remained just that—dreams. Through perseverance, self-discipline, and a clear plan, Laura transformed her life and achieved what she once thought was impossible. This narrative serves as a powerful reminder that action is the bridge between intention and transformation.

Clarity of Vision: The First Step Toward Change

Creating lasting change begins with having a clear vision of what you want to achieve. Without a well-defined goal, it's easy to become distracted, disheartened, or overwhelmed by the obstacles that arise. Whether the goal is personal growth, career advancement, improved health, or building meaningful relationships, clarity is essential.

It is crucial to understand why a particular goal matters. This deep understanding of one's aspirations provides the intrinsic motivation necessary to keep moving forward, even when challenges arise. By spending time reflecting on one's desires, values, and motivations, individuals gain the clarity needed to take purposeful action. This process of self-discovery not only provides direction but also fosters a sense of purpose, making it easier to overcome setbacks along the way.

Take, for example, Michael, who had always dreamed of running his own business. However, for years, he had hesitated, unsure of where to start. One day, he took a step back and asked himself why he wanted to start a business in the first place. He realized that his primary goal was to create a company that would allow him to help others by offering high-quality services. This clarity of vision gave Michael the motivation to begin researching the market, identifying his target audience, and outlining a business plan. Armed with a clear sense of purpose, Michael found the courage to take his first steps toward entrepreneurship, ultimately creating a successful and meaningful business.

Setting SMART Goals: The Blueprint for Success

Once individuals have a clear vision, the next step is to break down that vision into actionable goals. The SMART framework—Specific, Measurable, Achievable, Relevant, and Time-bound—is an invaluable tool in this process. By ensuring that goals are well-defined and broken into smaller steps, individuals create a roadmap that guides them toward success.

1. Specific: A specific goal clearly defines what you want to achieve. Instead of simply saying, "I want to be healthier," a specific goal would be, "I want to lose 10 pounds in three months by exercising 3 times a week and eating a balanced diet."
2. Measurable: A measurable goal includes criteria for tracking progress. For example, if your goal is to improve your writing skills, a measurable goal could be, "Write 500 words every day for the next month."
3. Achievable: Goals should be realistic. It's important to set goals that challenge you but are also within your capabilities. For instance, if you've never run before, aiming to run a marathon within a month is likely too ambitious. A more achievable goal might be, "Run 5 kilometers in three months."
4. Relevant: Goals should align with your broader aspirations and values. If your goal is to work in a creative field, pursuing a degree in fine arts might be more relevant than studying engineering.
5. Time-bound: Every goal should have a deadline. This creates urgency and helps you stay focused. For example, "Finish

    reading a self-improvement book within the next two weeks" is a time-bound goal.

The Importance of Accountability

While clarity, planning, and action are crucial elements of change, accountability is another important factor in creating lasting transformation. It's easy to let goals slip or become distracted, especially when faced with obstacles or competing priorities. That's where accountability comes in. Whether through self-monitoring, an accountability partner, or a supportive community, accountability helps individuals stay committed to their goals.

Having an accountability partner can significantly increase the likelihood of success. This person can provide motivation, offer advice, and hold you responsible for taking the necessary actions. Additionally, joining a group or community with similar goals can provide encouragement and feedback, reinforcing the importance of progress.

Overcoming Obstacles and Building Resilience

Throughout the process of creating lasting change, obstacles are inevitable. Life's unpredictability can throw curveballs, and setbacks are often a part of the journey. However, the ability to overcome these challenges and build resilience is what separates those who achieve lasting transformation from those who give up.

One of the most important qualities to cultivate is persistence. When faced with obstacles, successful individuals do not give up—they pivot, adjust their approach, and keep going. This mindset is what enables them to turn adversity into an opportunity for growth.

Conclusion: Creating Lasting Change

In conclusion, lasting change is not a result of a single moment of inspiration but a continuous journey of intentional action. It starts with a clear vision, followed by the setting of SMART goals that serve as a roadmap. Accountability and resilience play a critical role in maintaining momentum, while persistence ensures that individuals stay on course despite setbacks.

As you embark on your journey of transformation, remember that change takes time, effort, and patience. By taking actionable steps, embracing challenges, and remaining committed to your goals, you can create lasting change in your life. The journey may be long, but the rewards—personal growth, fulfillment, and achievement—are worth every step.

Chapter 10: The Role of Habit Formation in Change

The pursuit of meaningful, lasting change is a journey often shaped by small, consistent actions rather than dramatic shifts. One of the most effective ways to create sustained transformation in life is by cultivating positive habits. As the saying goes, "Success is the sum of small efforts, repeated day in and day out." In this chapter, we explore how the power of habit formation can be harnessed to foster positive behaviors and achieve long-term goals.

The Habit Loop: Cue, Routine, and Reward

To understand how habits are formed and maintained, it's essential to first explore the "habit loop." This concept, popularized by Charles Duhigg in his book *The Power of Habit*, explains how habits are constructed in three distinct stages:

1. Cue (Trigger): The habit loop begins with a cue, which is a trigger or a signal that prompts the brain to initiate a habitual behavior. This cue can be external, such as an alarm going off, or internal, such as a feeling of stress or hunger.
2. Routine: The routine is the behavior or action that follows the cue. It is the habitual action that we automatically perform in response to the trigger. For example, if the cue is stress, the routine might involve eating a snack or taking a walk.
3. Reward: After completing the routine, a reward is provided, reinforcing the behavior. The reward can be intrinsic, such as the feeling of satisfaction after exercising, or extrinsic, like receiving praise for completing a task. The reward strengthens

    the connection between the cue and the routine, making the behavior more likely to be repeated in the future.

Consider the story of David, who struggled with maintaining a healthy diet. His cue was often stress from work, which led him to crave junk food as a form of emotional comfort. The routine was eating chips or cookies in the evening, and the reward was the temporary relief from stress and a sense of enjoyment from the food.

David recognized this pattern and decided to change his habit loop. Instead of eating junk food when stressed, he replaced it with a healthier option—eating fruit or practicing mindful breathing exercises. Over time, the new routine provided a similar reward, but with a healthier outcome. The more David repeated this loop, the more ingrained the new habit became.

Understanding this loop is key to breaking bad habits and cultivating good ones. By identifying the cues that trigger unhealthy behaviors and replacing the routines with more beneficial actions, individuals can take control of their habits and align them with their long-term goals.

Building New Habits

The process of building new habits can be overwhelming if approached all at once. The key to success is starting small and scaling gradually. Trying to make drastic changes too quickly often leads to burnout or frustration, which can derail progress. Instead, begin with a habit that is achievable and easy to incorporate into your daily routine.

One effective strategy is to focus on the "two-minute rule," a concept introduced by James Clear in his book *Atomic Habits*. The two-minute rule states that any new habit should take no more than two minutes to complete. This approach helps reduce resistance to starting, as even the most daunting tasks can be broken down into a manageable chunk. Once the habit is ingrained, you can gradually increase the time and complexity of the action.

Self-Reflection and Adjustments in Habit Formation

Self-reflection plays a vital role in habit formation. As you work on building new habits, it's important to regularly assess your progress and make necessary adjustments. This reflection allows you to identify what's working well, what challenges you're facing, and where you might need to tweak your approach.

For example, if you're trying to build a habit of daily reading but find that evenings are too chaotic to sit down with a book, you could reflect on this and shift your reading time to the mornings or during lunch breaks. Reflection enables you to stay flexible and adjust your approach to suit your lifestyle and goals.

Self-Reflection:

Liam had set a goal of practicing mindfulness for 10 minutes every day, but after a few weeks, he realized that his initial morning routine was not working for him. He felt rushed in the mornings and often skipped his mindfulness session. After reflecting on his schedule, he decided to move his practice to the evening when he felt calmer and more focused. This change allowed Liam to maintain his mindfulness practice, and over time, it became a natural part of his nightly routine.

Self-reflection allows for continuous improvement in the habit-building process, ensuring that habits are sustainable and effective in achieving desired outcomes.

The Role of Motivation and Accountability in Habit Formation

While the science of habit formation provides valuable insights, maintaining motivation and accountability are also critical for success. Motivation is often highest at the beginning of a new habit, but it can dwindle over time. To counter this, it's helpful to create external systems of support, such as an accountability partner or a community of individuals with similar goals.

Accountability helps individuals stay committed to their habits, especially during times when motivation is low. Whether through regular check-ins with a friend or logging progress in a habit-tracking app, these external structures create a sense of responsibility that encourages consistency.

Motivation and Accountability:

Sophie, a college student, wanted to build the habit of studying for at least one hour every day. To stay motivated, she partnered with a classmate and agreed to check in with each other daily. They would send a quick message confirming whether they had completed their study session for the day. This simple accountability system helped Sophie stay consistent and motivated, even when her workload became heavy.

In addition to accountability, maintaining intrinsic motivation is key. Individuals should remind themselves of why they want to form a particular habit and the long-term benefits it will bring. Reflecting on the positive impact of their new habits reinforces commitment and keeps individuals focused on their goals.

THE MIND ART

Consistency Over Perfection

One of the most important aspects of habit formation is understanding that consistency is more important than perfection. It's easy to get discouraged when we miss a day or fail to meet a goal, but setbacks are a natural part of the process. The key is to get back on track and keep going, rather than letting one mistake derail your progress.

Small, consistent actions over time yield the most significant results. By focusing on the process of habit-building rather than striving for perfection, individuals create a sustainable path to personal growth.

Empowering Change Through Habits

In conclusion, habit formation is a powerful tool for creating lasting change in our lives. By understanding the habit loop—cue, routine, and reward—and using strategies like starting small, self-reflection, and accountability, individuals can successfully build new habits that align with their goals and aspirations.

The journey toward personal transformation is not about making drastic changes overnight, but rather about creating a foundation of consistent, positive habits that compound over time. Whether it's improving health, enhancing productivity, or fostering relationships, habits are the building blocks of meaningful, lasting change. By committing to the process and embracing consistency, individuals can shape a life that reflects their true values and desires.

LAHARI KORUTLA
Chapter 11: The Journey of Lifelong Learning

In a world that is constantly evolving, the pursuit of knowledge and personal growth is no longer confined to the classroom or a specific phase of life. Lifelong learning has become an essential part of adapting to change and remaining relevant, whether in personal or professional contexts. This chapter explores the importance of continuous learning, how it enriches our lives, and practical strategies for making it a part of everyday existence. As Albert Einstein wisely said, *"Intellectual growth should commence at birth and cease only at death,"* reinforcing the idea that learning is a lifelong journey.

The Importance of Lifelong Learning

At its core, lifelong learning is about cultivating a mindset that seeks knowledge and growth at every stage of life. It's a commitment to embracing curiosity, exploring new ideas, and developing new skills. In a fast-paced, ever-changing world, those who engage in lifelong learning are better equipped to navigate the complexities of modern life. They remain adaptable, resilient, and open to new opportunities, enabling them to thrive in an uncertain future.

Lifelong learning also fosters intellectual growth, which in turn enhances emotional and psychological well-being. Engaging in learning activities can stimulate the brain, improve memory, and reduce the effects of aging on cognitive function. Moreover, when individuals engage in meaningful learning experiences, they build confidence and a sense of accomplishment that extends beyond the acquisition of knowledge.

Consider the story of Lisa, a woman in her 40s who decided to learn a new language after years of focusing on her career and raising a family. Initially, Lisa felt intimidated by the idea of learning something entirely new. However, she embraced the challenge with a mindset of curiosity and perseverance. Over time, Lisa not only learned to speak Spanish fluently but also discovered a new sense of confidence in her ability to acquire new skills. This journey of learning reshaped her perspective, reminding her that it's never too late to learn and grow.

Adaptability and Resilience Through Learning

In an era defined by rapid technological advancements, political shifts, and global challenges, adaptability is one of the most critical skills an individual can possess. Lifelong learning plays a crucial role in developing this skill. By continuously learning, individuals can better adapt to changes in their personal and professional environments. They stay informed, adjust to new technologies, and remain competitive in a constantly evolving world.

Resilience, too, is cultivated through lifelong learning. When individuals commit to learning, they develop the mental agility to handle setbacks, failures, and obstacles with a growth mindset. Instead of viewing challenges as insurmountable, they see them as opportunities for growth and improvement. This ability to adapt, learn from experiences, and bounce back from adversity is an essential component of resilience.

Take the story of John, a graphic designer who found himself out of work after his company downsized. Initially, he was devastated and unsure of his next steps. However, John decided to embrace learning as a way to navigate his new reality. He took online courses in web development, expanding his skill set beyond design. This not only made him more marketable but also gave him the resilience to face the challenges of unemployment with optimism. After several months of learning, John landed a new job as a web designer, combining his graphic design expertise with his new web development skills. Through lifelong learning, John transformed a setback into a springboard for career growth.

Strategies for Lifelong Learning

Incorporating lifelong learning into everyday life doesn't require a major overhaul of one's routine. Instead, it involves small, intentional steps that create opportunities for continuous growth. Here are a few practical strategies to help integrate learning into daily life:

1. Set Aside Time for Reading: Reading remains one of the most effective ways to gain knowledge and expand one's perspective. Setting aside time for reading each day can have a significant impact on intellectual growth. Whether it's reading books, articles, or research papers, dedicating time to consume new information fosters continuous learning. Start small by setting a goal of reading just 15-20 minutes per day and gradually increase the time as it becomes a habit.
2. Engage in Online Courses: With the rise of online platforms, learning has become more accessible than ever before. Websites like Coursera, Udemy, and Khan Academy offer courses on a wide range of topics, from technical skills like coding and data analysis to personal development and creative writing. By enrolling in an online course, individuals can learn at their own pace and acquire new skills that align with their goals.
3. Attend Workshops and Seminars: Attending workshops and seminars provides opportunities for hands-on learning and real-time interaction with experts in various fields. These events allow individuals to expand their knowledge base, build connections with others, and gain valuable insights that may not be available through self-paced learning alone.
4. Explore New Hobbies and Interests: Lifelong learning doesn't have to be strictly academic or professional. Exploring new hobbies and interests, whether it's painting, gardening, photography, or learning to play a musical instrument, stimulates the mind and keeps it active. These pursuits can lead to a deeper understanding of oneself and the world around them.

5. **Reflect on What You've Learned:** Regular reflection on your learning journey is essential. It allows you to assess how the knowledge you've gained applies to your life and helps reinforce what you've learned. Reflecting on new insights and skills can also open doors for new learning opportunities and encourage personal growth.

Sophia, a marketing professional, wanted to broaden her skills by learning about artificial intelligence (AI) and its applications in marketing. She set aside 30 minutes each morning to read articles about AI. She also enrolled in an online course on AI for marketing and attended a local workshop about the future of technology in the industry. Over the course of several months, Sophia's understanding of AI grew, and she applied this knowledge to create more effective marketing strategies for her company. This commitment to continuous learning helped Sophia stay ahead of industry trends and advance in her career.

Real-Life Transformations Through Lifelong Learning

Real-life examples of individuals who have embraced lifelong learning highlight its transformative power. Consider the story of Mary, a retired teacher who, after years of educating students, decided to pursue her passion for painting. Initially, Mary lacked formal training, but she started taking painting classes at a local art school. Through consistent practice and learning, she developed a unique style and began exhibiting her work in galleries. Mary's journey shows that learning doesn't stop after retirement or when one reaches a certain age; instead, it can continue to open doors to new passions and opportunities.

Another inspiring example is the story of Michael, a high school dropout who struggled in his early years but eventually turned his life around by embracing education. After completing his GED, Michael enrolled in community college and then transferred to a university to study business. Through hard work and dedication, he eventually became an entrepreneur, founding a successful tech startup. Michael's story demonstrates that it's never too late to start learning and that education can be the key to unlocking new opportunities, regardless of where you start.

The Role of Reflection in Lifelong Learning

Reflection is a cornerstone of the lifelong learning process. It allows individuals to assess their progress, identify gaps in their knowledge, and refine their approach. By regularly asking questions like, *"What have I learned recently?"* and *"How can I apply this knowledge to my life?"*, individuals can ensure that their learning is purposeful and aligned with their personal and professional goals.

Reflection also allows learners to appreciate the journey of growth. It reminds them that learning is not a race to the finish line, but a continuous process that unfolds over time. When individuals reflect on their learning experiences, they develop a deeper understanding of themselves and their abilities, which fosters a sense of fulfillment and motivation to keep learning.

Empowering Personal Growth Through Lifelong Learning

In conclusion, lifelong learning is not just a way to acquire knowledge; it is a powerful tool for personal growth and development. By adopting a mindset of curiosity and commitment to continuous education, individuals can remain adaptable, resilient, and open to new possibilities. Lifelong learning empowers individuals to stay relevant, face challenges with confidence, and unlock their full potential. In a world filled with constant change, the ability to learn and grow is the greatest asset one can have. Therefore, embracing lifelong learning is not just an option but a vital strategy for thriving in an ever-changing world.

THE MIND ART

Chapter 12: The Science of Motivation

Motivation is the force that propels us toward our goals and dreams. However, it is not a single, uniform entity; it is a complex and deeply personal experience influenced by various internal and external factors. Understanding what motivates us is crucial to unlocking our potential and driving consistent progress in our lives. This chapter delves into the science of motivation, exploring the underlying theories, offering practical strategies for enhancing motivation, and illustrating how understanding one's personal drive can lead to transformative success.

The Nature of Motivation

At the heart of motivation lies a fundamental question: *What drives us?* For some, motivation may stem from intrinsic factors, such as a deep sense of purpose, personal satisfaction, or the joy of mastery. For others, it might be extrinsic factors, such as rewards, recognition, or the desire to meet external expectations. In many cases, a combination of both intrinsic and extrinsic motivators plays a role in shaping behavior.

A powerful example of intrinsic motivation can be seen in the case of someone pursuing their passion—let's say, an artist dedicating hours to painting because they love the creative process and the personal satisfaction it brings. On the other hand, extrinsic motivation might come from a desire to earn a promotion at work, win an award, or achieve recognition from others. Understanding how both intrinsic and extrinsic motivators work together is crucial to sustaining long-term motivation.

Personal Motivation:

Consider the story of Sarah, a writer who had always dreamed of publishing a book. Despite numerous setbacks and rejections, her intrinsic motivation—her passion for storytelling—kept her going. At times, the external validation she sought (like positive feedback or recognition from literary agents) also fueled her drive. But ultimately, it was her unwavering passion for writing that pushed her to keep honing her craft. Over time, Sarah successfully published her first novel, which went on to receive critical acclaim. Her story illustrates how motivation is a dynamic force shaped by both internal passions and external factors.

THE MIND ART
Maslow's Hierarchy of Needs and Motivation

One of the most widely known frameworks for understanding human motivation is Maslow's Hierarchy of Needs, proposed by psychologist Abraham Maslow. This theory suggests that human needs are arranged in a five-tier pyramid, starting with the most basic physiological needs and moving toward self-actualization, which represents the realization of one's fullest potential.

1. Physiological Needs: Basic needs like food, water, shelter, and sleep are fundamental for survival. Until these needs are met, higher-level motivations cannot fully take root.
2. Safety Needs: Once physiological needs are met, the need for security, stability, and protection arises.
3. Social Needs: These include the need for relationships, love, and belongingness, such as friendships, family, and social connections.
4. Esteem Needs: After social needs are met, individuals seek respect, recognition, and a sense of achievement.
5. Self-Actualization: The pinnacle of Maslow's hierarchy, selfactualization, is the desire to become the best version of oneself, reaching one's fullest potential and finding personal fulfillment.

LAHARI KORUTLA

According to Maslow, people are motivated to fulfill their most basic needs before they can pursue higher-level aspirations. When basic needs are not met, motivation tends to be focused on securing those needs first. As individuals' needs are progressively satisfied, their focus shifts toward more abstract, personal goals—such as self-improvement, creativity, and personal growth.

Maslow's Hierarchy in Action:

Take the example of Amir, a young entrepreneur who grew up in a low-income household. Early in his life, his motivation was primarily focused on securing basic needs, such as food and shelter. Once Amir's financial situation stabilized, he began pursuing his passion for technology, aiming for professional success and recognition. As his career flourished, his motivation shifted toward more profound goals—such as using his success to help others and make a positive impact on society. Amir's story demonstrates how motivation evolves as individual needs are met and a person moves up the hierarchy.

Self-Determination Theory and Motivation

Another influential framework for understanding motivation is Self-Determination Theory (SDT), which emphasizes the role of autonomy, competence, and relatedness in fostering intrinsic motivation. According to SDT, individuals are most motivated when they feel that they are in control of their actions (autonomy), when they feel competent or capable in their pursuits (competence), and when they feel connected to others (relatedness).

Autonomy refers to the sense of being in control of one's actions, rather than feeling compelled or pressured by external forces.

Competence involves the ability to effectively engage in and master tasks, fostering a sense of achievement.

Relatedness is the need for social connections and a sense of belonging.

When these three elements are present, individuals are more likely to experience intrinsic motivation, which leads to sustained engagement and satisfaction in their pursuits.

Self-Determination in Action:

Maria, a teacher, always felt more motivated to engage with her students when she had the freedom to design her curriculum and teaching methods. By tapping into her creativity (autonomy) and her passion for fostering student growth (competence), Maria found that her motivation to teach was not only sustained but flourished. Additionally, the relationships she built with her students (relatedness) further reinforced her intrinsic motivation. As a result, Maria experienced fulfillment and success, both personally and professionally.

These strategies help individuals harness their inner drive, enabling them to achieve their goals and dreams.

Goal-Setting: Setting clear, achievable goals is one of the most powerful ways to enhance motivation. Using the SMART (Specific, Measurable, Achievable, Relevant, Time-bound) framework helps break down large, daunting tasks into smaller, manageable steps. This approach boosts confidence and provides a sense of accomplishment along the way.

Visualization: Visualization is a technique that involves mentally picturing oneself achieving a goal. By vividly imagining success, individuals increase their belief in their abilities, which boosts motivation. This strategy can be particularly effective for athletes, performers, and anyone pursuing a challenging goal.

Reward Systems: Establishing reward systems is another effective strategy to maintain motivation. By rewarding progress, even small achievements, individuals reinforce positive behaviors. For instance, a person pursuing fitness goals might reward themselves with a treat or a fun activity after completing a week of workouts.

Find Your Why: Often, motivation falters when the underlying reasons for pursuing a goal become unclear. Finding a deep, personal reason (your "why") for your goals can serve as a powerful motivator. This deep connection to the purpose behind your actions makes it easier to stay focused and committed, especially during times of difficulty.

Goal-Setting and Motivation:

Jake, an aspiring musician, had always dreamed of composing his own album. However, despite his talent, he struggled with motivation and consistency. By breaking down the larger goal into smaller, achievable tasks (such as writing one song per month), Jake was able to maintain momentum. He also set up a reward system, treating himself to a special dinner after each milestone. Over time, Jake not only completed his album but also found renewed inspiration and passion for his craft. His story demonstrates the power of goal-setting and reward systems in fueling motivation.

The Role of Resilience in Motivation

Even with the best strategies in place, setbacks and challenges are inevitable. Resilience—the ability to bounce back after adversity—is a critical factor in sustaining motivation. When faced with obstacles, resilient individuals are more likely to persevere and continue pursuing their goals, rather than giving up.

Building resilience involves developing coping mechanisms for dealing with failure, maintaining a positive mindset, and seeking support when needed. By fostering resilience, individuals can maintain their motivation through difficult times and continue on their path to success.

Understanding and Cultivating Motivation

In conclusion, motivation is a dynamic, multifaceted force that drives individuals to pursue their dreams and aspirations. By understanding the science behind motivation, including theories like Maslow's Hierarchy of Needs and Self-Determination Theory, individuals can better understand their personal drivers and how to sustain their motivation over time. Whether through goal-setting, visualization, or fostering resilience, motivation is a tool that can be cultivated and refined. By understanding what truly motivates us and aligning our actions with our values, we can achieve our goals and create lasting change in our lives.

Chapter 13: The Art of Communication

Communication is a foundational skill that shapes every aspect of our lives. From personal relationships to professional success, the ability to communicate effectively is vital. The story of an individual who transformed their relationships through enhanced communication skills underscores the power of dialogue. Effective communication is not just about speaking clearly; it is about listening with the intent to understand, expressing thoughts clearly, and navigating emotions with empathy. This chapter explores the art of communication, delving into its core components, offering practical strategies for improvement, and highlighting how mastering communication can lead to more meaningful connections and personal growth.

The Essence of Effective Communication

Effective communication is often misunderstood as simply talking or exchanging information. However, communication is a dynamic process that involves both the transmission and reception of messages, shaped by verbal and non-verbal cues. As Stephen R. Covey aptly noted, "Most people do not listen with the intent to understand; they listen with the intent to reply." This quote highlights a crucial aspect of communication: *listening*.

THE MIND ART

Communication isn't just about expressing your thoughts; it is about understanding the perspectives and emotions of others.

At its core, effective communication is a two-way process involving not only the speaker but also the listener. While verbal communication conveys the message through words, non-verbal communication—such as body language, tone of voice, and facial expressions—plays an equally important role in how messages are received and understood.

For example, consider a meeting between a manager and an employee. The manager may say all the right words, but if their body language conveys disinterest or impatience, the employee may feel undervalued or dismissed. Effective communication requires consistency between what is said and what is expressed non-verbally.

Active Listening: The Key to Connection

One of the most powerful tools in effective communication is active listening. Active listening goes beyond hearing words; it requires full engagement with the speaker, where the listener gives their undivided attention, absorbs the information, and responds thoughtfully. Active listening involves three key elements: paying attention, showing that you are listening, and providing feedback.

paying Attention: Active listening begins with focusing on the speaker. This means setting aside distractions, such as phones or multitasking, and giving the speaker your full attention.

Showing That You Are Listening: Non-verbal cues such as nodding, maintaining eye contact, and using facial expressions help demonstrate attentiveness.

Providing Feedback: Responding to the speaker with paraphrasing or asking clarifying questions ensures that the listener has understood the message and creates an opportunity for further clarification if needed.

Active Listening:

Consider a conversation between a husband and wife after a long day. The wife expresses her frustration about a difficult situation at work. An active listener would focus on her words, maintain eye contact, and nod occasionally to show empathy. After she finishes, the listener might say, "It sounds like you felt overwhelmed by the expectations at work today. How can I help?" This feedback demonstrates both understanding and support, creating an environment where the wife feels heard and valued.

Active listening not only improves the flow of information but also strengthens relationships. It shows respect and empathy, building trust and creating a deeper sense of connection between individuals.

Verbal and Non-Verbal Communication

Communication is not just about words; how we say things is just as important. Verbal communication encompasses the language, tone, and structure of what is being said. Words carry meaning, but tone and delivery influence how those words are interpreted. For example, the statement "I'm fine" can be said in many ways, but depending on the tone, it might signal frustration, indifference, or genuine well-being.

Non-verbal communication, on the other hand, refers to body language, gestures, posture, facial expressions, and eye contact. These non-verbal cues often reveal more about a person's true feelings than the words they use. For instance, crossed arms may suggest defensiveness or discomfort, while open posture can convey receptiveness and warmth.

Non-Verbal Cues in Communication:

A manager addressing a team might use open gestures, a calm tone, and positive facial expressions to convey confidence and encouragement. Conversely, if the manager's arms are crossed, their voice is tense, and their gaze is avoidant, the team may interpret the communication as unapproachable or disengaged. Non-verbal cues are powerful tools in communication because they often express emotions and attitudes that words cannot fully capture.

The key to mastering communication is aligning verbal and non-verbal signals. When these elements are in harmony, the message is clear and effective. Discrepancies between verbal and non-verbal communication can create confusion and undermine the message's intent.

The Heart of Effective Communication

Empathy plays a central role in communication. It allows individuals to step into the shoes of others, understand their feelings, and respond in ways that acknowledge their emotions.

Empathy is essential in both personal and professional relationships, as it fosters trust, reduces conflict, and strengthens connections.

Effective communicators are not only aware of their own emotions but are attuned to the emotional states of others. By acknowledging and validating others' feelings, individuals create an environment where everyone feels respected and heard.

Empathy in Action:

Imagine a colleague who is going through a tough time in their personal life and expresses feelings of overwhelm at work. An empathetic response would involve listening without judgment, acknowledging their feelings, and offering support. Instead of simply offering a solution, an empathetic communicator might say, "I can see that you're feeling a lot right now. How can I help?" This response shows understanding and care, which can provide the colleague with the emotional support they need to navigate the situation.

Empathy also helps navigate conflicts. When individuals feel heard and understood, they are more likely to approach problem-solving collaboratively rather than defensively. This strengthens the communication process and enhances relationships.

Navigating Difficult Conversations

Difficult conversations are an inevitable part of life, whether in personal relationships or the workplace. Handling them with grace and tact requires preparation and skill. The chapter introduces several strategies for navigating these challenging dialogues:

Stay Calm and Focused: Maintaining composure is critical when emotions are running high. Staying calm allows you to think clearly and respond thoughtfully rather than react impulsively.

Use "I" Statements: Instead of placing blame, use "I" statements to express how you feel. For example, "I feel frustrated when..." instead of
"You always...".

Listen and Acknowledge: Even in difficult conversations, active listening and empathy go a long way in de-escalating tension.

Find Common Ground: Seek areas of agreement and work from there. Finding common ground fosters cooperation and reduces adversarial feelings.

Handling a Difficult Conversation:

Consider a situation where a team member feels frustrated with the performance of their colleague. Instead of confronting them aggressively, an effective communicator might initiate the conversation by saying, "I've noticed some challenges with our team's progress, and I'd like to understand your perspective on it. How can we work together to improve things?" This approach focuses on collaboration rather than blame, increasing the chances of a positive outcome.

Emotional Intelligence in Communication

Emotional intelligence (EQ) is a critical factor in effective communication. It involves the ability to recognize and manage your emotions and the emotions of others. High emotional intelligence allows individuals to remain calm and empathetic in the face of challenges, fostering better communication and more successful outcomes in interpersonal interactions.

Emotional intelligence enhances self-awareness, self-regulation, motivation, empathy, and social skills. By improving these skills, individuals can navigate complex emotions and create more harmonious relationships.

Emotional Intelligence in Communication:

A leader with high emotional intelligence might notice that a team member is feeling anxious before a presentation. Instead of focusing solely on the content, the leader might offer encouragement and reassurance: "I know you've been preparing hard for this, and I believe you're ready. If you need anything, I'm here for support." This approach not only addresses the immediate concern but also builds trust and confidence within the team.

The Power of Effective Communication

Effective communication is a skill that can be developed and refined with practice. It requires active listening, empathy, clarity, and emotional intelligence. By mastering these components, individuals can enhance their relationships, foster collaboration, resolve conflicts, and navigate difficult conversations with grace and confidence. Whether in personal relationships or the workplace, communication is the bridge that connects us to others, allowing us to share ideas, emotions, and experiences.

The journey of mastering communication is ongoing. By committing to improvement and applying these principles, individuals can become more powerful communicators, ultimately enriching their lives and the lives of those around them.

Chapter 14: The Influence of Self-Talk

Self-talk, the internal dialogue that runs in our minds, holds immense power in shaping our perceptions, emotions, and behaviors. It influences how we approach challenges, view our capabilities, and ultimately impact our success in life. As James Allen famously stated, "As a man thinketh in his heart, so is he," reminding us of the profound connection between our thoughts and the reality we experience. This chapter explores the impact of self-talk, illustrating how changing our inner dialogue can transform our mindset and, consequently, our lives.

Understanding the Power of Self-Talk

Self-talk refers to the ongoing conversation we have with ourselves, often without conscious awareness. It can be positive, where we encourage and uplift ourselves, or negative, where we criticize and diminish our worth. Our self-talk plays a critical role in shaping how we view ourselves and how we react to external circumstances. It influences everything from our confidence and self-esteem to our approach to challenges and setbacks.

When we engage in negative self-talk, we may constantly tell ourselves that we are not good enough, capable enough, or deserving of success. These thoughts create feelings of inadequacy, anxiety, and fear of failure, which can prevent us from pursuing opportunities or taking risks. On the other hand, positive self-talk encourages self-belief, optimism, and resilience, empowering us to move forward with confidence, even in the face of adversity.

For instance, imagine someone preparing for a job interview. If their inner dialogue is negative, they may think, "I'm not qualified for this position," or "I'm going to mess up," which may lead to nervousness and a lack of preparedness. Alternatively, if their self-talk is positive, they might think, "I have the skills for this role," or "I will do my best," which boosts their confidence and enhances their performance.

The Science Behind Self-Talk

The psychology behind self-talk is rooted in cognitive psychology, which suggests that our thoughts significantly influence our emotions and behaviors. According to cognitive-behavioral theory, negative thought patterns contribute to negative emotional responses and, ultimately, maladaptive behaviors. For example, when we engage in self-criticism, we tend to feel overwhelmed, discouraged, and less motivated to take action.

Conversely, positive self-talk helps reframe challenges and setbacks as opportunities for growth. By changing the way we speak to ourselves, we can shift our emotional responses and behavior, leading to healthier coping mechanisms and more adaptive actions. This transformation is not instant but requires consistent practice and awareness of our inner dialogue.

Studies in psychology also support the idea that self-talk can influence performance. Research has shown that athletes who use positive self-talk experience improved performance, greater focus, and reduced anxiety. This principle can be applied in everyday life, from academics to relationships, as a way of boosting our confidence and maintaining a positive outlook, even when faced with difficulties.

Positive vs. Negative Self-Talk: How They Shape Our Lives

The distinction between positive and negative self-talk is critical. Positive self-talk is characterized by affirmations, encouragement, and optimism. It involves recognizing strengths and focusing on solutions rather than problems. Negative self-talk, on the other hand, is filled with self-doubt, criticism, and fear. It often manifests as harsh judgments or unrealistic expectations, such as "I'm not good enough" or
"I always fail."

Example of Positive Self-Talk:

Consider a student preparing for a final exam. If they engage in positive self-talk, they might say, "I have studied hard, and I am prepared. I can handle this," which cultivates confidence and reduces stress. This mindset helps them stay focused during the exam, improving their chances of performing well.

On the other hand, if the student's self-talk is negative, they may think, "I'm going to fail. I can't remember anything," leading to anxiety and distraction. This negative self-talk undermines their ability to perform at their best, creating a self-fulfilling prophecy.

The key to changing our outcomes often lies in changing our internal narrative. Shifting from a mindset of limitation to one of possibility enables us to confront challenges with resilience and take actionable steps toward our goals.

Cultivating Positive Self-Talk: Practical Strategies

Transforming self-talk requires intentional effort. It is not enough to simply wish for a more positive mindset; it requires actionable strategies to make lasting change. Here are some practical techniques for cultivating positive self-talk:

Affirmations: One of the simplest and most effective ways to shift self-talk is through the use of positive affirmations. These are statements that reinforce a positive belief about yourself or your capabilities. For example, "I am capable and deserving of success," or "I am confident and resilient." Repeating affirmations regularly helps reprogram the mind and replace negative thoughts with empowering beliefs.

Cognitive Restructuring: Cognitive restructuring is a technique from cognitive-behavioral therapy that involves identifying negative thought patterns and replacing them with more balanced or positive alternatives. For example, if you catch yourself thinking, "I always mess up," challenge that thought by asking, "Is that really true? Have I succeeded in similar situations before?" By reframing negative thoughts, you can transform your internal narrative and improve your emotional state.

Gratitude Practice: Focusing on gratitude shifts the mind away from negativity and scarcity toward positivity and abundance. Taking time each day to acknowledge the things you are grateful for can foster a more optimistic outlook on life, reducing the prevalence of negative self-talk.

Visualization: Visualization involves mentally rehearsing positive outcomes or successful scenarios. By picturing yourself succeeding in your goals, you reinforce a belief in your ability to achieve them. This practice not only enhances confidence but also primes the brain to focus on solutions and success.

Mindfulness and Self-Awareness: Mindfulness practices help individuals become more aware of their thoughts and feelings in the present moment. By cultivating this awareness, you can catch negative self-talk before it spirals out of control and replace it with more constructive thoughts.

THE MIND ART

Self-Compassion and Its Role in Self-Talk

An important aspect of self-talk that is often overlooked is self-compassion. Self-compassion involves treating oneself with kindness and understanding, especially during times of failure or struggle. Instead of harshly criticizing ourselves, self-compassion allows us to recognize our mistakes without judgment and encourages a nurturing attitude toward our own well-being.

When we practice self-compassion, we are more likely to engage in positive self-talk. For instance, instead of berating ourselves for making a mistake, we can say, "It's okay to make mistakes. I'm learning and growing from this experience." This approach reduces the impact of negative self-talk, promoting emotional well-being and resilience.

Example of Self-Compassion in Action:

Consider an athlete who misses an important shot during a game. Instead of saying, "I'm terrible, I let the team down," they might practice self-compassion by acknowledging the mistake but reframing it: "I missed that shot, but that doesn't define me. I'll keep practicing and get better next time." This compassionate self-talk helps them bounce back quickly, maintaining a positive mindset for the remainder of the game.

The Transformative Power of Self-Talk

Self-talk is a powerful tool that shapes how we view ourselves, our challenges, and our potential. By becoming aware of our internal dialogue and intentionally shifting from negative to positive self-talk, we can unlock our full potential and create a more fulfilling life. The strategies outlined in this chapter—from affirmations to cognitive restructuring—are not quick fixes but practices that require consistent effort and awareness.

The stories of individuals who have transformed their lives through positive self-talk illustrate that change is possible. With dedication and the right tools, we can reframe our internal narrative and embrace a mindset that supports our growth and success. As we continue to nurture positive self-talk and practice self-compassion, we empower ourselves to face life's challenges with confidence, resilience, and the belief that we are capable of achieving our dreams.

LAHARI KORUTLA
Chapter 15: Finding Your Purpose

Finding one's purpose is a transformative journey that leads to a deep sense of meaning and fulfillment. It is a quest that requires introspection, self-awareness, and a willingness to explore one's values, passions, and aspirations. As Viktor Frankl aptly stated, "Life is never made unbearable by circumstances, but only by lack of meaning and purpose." This powerful quote underscores the profound effect that having a clear sense of purpose can have on overall well-being. When we understand our purpose, we can navigate life's challenges with resilience and find satisfaction in both our personal and professional lives.

The Search for Purpose: A Personal Journey

The search for purpose often begins as a quiet whisper within, a feeling of longing for something more meaningful. Many people experience moments in life where they feel disconnected from their true selves or uncertain about their direction. This is part of the natural process of self-discovery. However, finding one's purpose does not happen overnight. It is a journey that unfolds gradually through self-reflection, experiences, and personal growth.

A powerful example of this journey can be seen in the life of Sarah, a successful corporate lawyer who, despite achieving external markers of success, felt unfulfilled. She spent years working in a field that didn't align with her true passions and values. Through a period of self-reflection and exploration, Sarah realized her true purpose was to help underprivileged children gain access to education. She left her high-paying job and began working with non-profit organizations to improve educational opportunities in underserved communities. This decision not only gave her a renewed sense of purpose but also a deeper sense of fulfillment. Sarah's story illustrates that purpose can often be discovered through a combination of introspection and action.

Reflecting on Values, Passions, and Aspirations

To embark on the journey of finding your purpose, it's essential to begin with an honest assessment of your values, passions, and aspirations. These elements serve as a compass, guiding individuals toward what truly matters to them. Understanding one's values—such as integrity, compassion, creativity, or justice—helps to clarify what is important in life and what one should prioritize. Similarly, identifying passions—the activities or causes that bring joy and fulfillment—can serve as a foundation for discovering a sense of purpose.

Practical exercises to guide this reflection include journaling prompts, such as:

- *What activities make me lose track of time?*
- *When have I felt most alive or inspired?*
- *What values do I hold most dear in my life?*

These reflective questions can provide insight into what brings meaning to an individual's life. Sometimes, purpose may not be immediately clear, but these exercises can help uncover patterns and areas of life where fulfillment is most often found. This process of introspection allows individuals to get in touch with their authentic selves and lay the groundwork for aligning their actions with their core values.

Example of Purpose Discovery:

Take the case of James, an artist who spent years in a corporate job that he disliked. He had always been passionate about painting but felt he could never make a living doing so. After a period of deep reflection and soul-searching, James realized that his passion for art was not just a hobby—it was his purpose. He began dedicating his time to painting full-time and also started teaching art to children in his community. His work became a source of joy not only for himself but for those around him as well. James' journey of discovering his purpose illustrates how passions, when nurtured and given space to grow, can lead to a fulfilling life.

Aligning Actions with Core Values

Once an individual has a clearer sense of their purpose, it's essential to align their actions with their core values. Purpose-driven living involves setting intentions that reflect the deeper motivations and desires that align with one's authentic self. The gap between knowing what one wants and actively pursuing it often lies in setting clear and purposeful goals.

For instance, if an individual values creativity and self-expression, their purpose might revolve around artistic endeavors, entrepreneurship, or working in a field that allows for creative problem-solving. Aligning actions with these values could involve setting specific goals like "launch a creative business" or "write and publish a book." These goals provide direction and motivation, as they resonate with an individual's deeper sense of self.

To effectively pursue a purpose-driven life, individuals must also stay committed to their goals, even when faced with challenges. Having a clear roadmap for achieving these goals—one that is in harmony with personal values—helps to maintain focus and momentum.

Example of Aligning Actions with Values:

Emma, a woman passionate about environmental sustainability, felt disconnected from her corporate job in marketing. After reflecting on her core values and interests, she recognized that her true purpose lay in advocating for a cleaner environment. She took concrete steps to align her actions with her values by transitioning into a career in environmental advocacy. Emma set specific goals, such as working with eco-conscious brands and promoting sustainability initiatives. She also volunteered for local environmental organizations to build experience and connections in the field. By aligning her professional path with her values, Emma found a sense of fulfillment that had been lacking in her previous job.

Setting Purpose-Driven Goals

Goal-setting is a critical aspect of finding and fulfilling one's purpose. Once individuals have clarified their purpose, they can begin setting goals that are both achievable and meaningful. These goals should be aligned with personal values, passions, and long-term aspirations. Purpose-driven goals are not merely about achieving external markers of success; they are about making progress toward a life that reflects an individual's truest self.

Practical strategies for setting purpose-driven goals include:

Creating SMART Goals: Break down large, overarching purposes into smaller, actionable steps using the SMART framework—Specific, Measurable, Achievable, Relevant, and Time-bound.

Identifying Milestones: Set intermediate milestones that reflect progress toward the larger purpose. These milestones help to stay motivated and provide a sense of accomplishment along the way.

Reviewing and Adjusting Goals: Periodically reassess goals and ensure they remain aligned with your evolving sense of purpose. Flexibility is essential, as your understanding of your purpose may deepen over time.

THE MIND ART

The Journey of Self-Discovery: Real-Life Inspiration

Real-life stories of individuals who have found their purpose offer valuable lessons and inspiration. Their stories highlight that purpose can be discovered through exploration, reflection, and sometimes even challenges. These narratives show that finding one's purpose is not always linear; it may evolve through different phases of life and can often arise from unexpected sources.

For example, Steve Jobs, the co-founder of Apple, experienced multiple setbacks in his life before discovering his true calling. He was fired from the company he founded, only to return later and revolutionize technology. Jobs once said, "Your work is going to fill a large part of your life, and the only way to be truly satisfied is to do what you believe is great work." His journey demonstrates that setbacks can often serve as stepping stones to finding one's true purpose.

Similarly, Malala Yousafzai's life purpose became clear after surviving a violent attack for advocating girls' education in Pakistan. Despite her traumatic experience, Malala's mission to promote education for girls around the world became her driving force. Through her activism, she has touched millions of lives and inspired a global movement for education.

Conclusion:

Embracing the Journey

In conclusion, finding one's purpose is an essential aspect of personal growth and fulfillment. It is a deeply personal and transformative process that requires time, introspection, and alignment with core values. By reflecting on values, passions, and aspirations, individuals can begin to uncover their true purpose and take intentional actions toward living a life of meaning. As individuals embark on this journey, they are empowered to create purpose-driven goals, maintain motivation, and stay committed to their path.

The process of discovering and living with purpose is not always easy, but it is one of the most rewarding aspects of personal development. The real-life examples of those who have found their purpose serve as inspiration, reminding us that purpose is not a destination but a lifelong journey. Embracing the path of self-discovery and aligning actions with values allows individuals to create a fulfilling and meaningful life—one that resonates deeply with their authentic selves.

Chapter 16: The Journey of SelfDiscovery

Self-discovery is often considered one of the most transformative and profound journeys in life. It involves an exploration of one's inner world, peeling back layers of external expectations and societal influences to reveal the true self. As Carl Jung wisely said, "The privilege of a lifetime is to become who you truly are," highlighting the significance of authenticity in the human experience. This journey is continuous, evolving, and deeply personal, as individuals uncover more about themselves with every step they take. In this chapter, we will delve into the process of self-discovery, providing guidance and real-life examples to inspire readers to embark on their own path to self-awareness.

The Importance of Self-Discovery

Self-discovery is not merely about uncovering hidden aspects of our personality or desires; it is about gaining clarity on who we are at our core. It involves understanding our values, beliefs, motivations, and what drives us toward certain goals and decisions. This process is deeply tied to our sense of purpose and fulfillment in life. When we truly understand ourselves, we are better equipped to navigate the challenges that life presents, make decisions that align with our true desires, and build relationships that nourish our well-being.

The journey of self-discovery is not a one-time event, but rather a lifelong process. Each stage of life brings new insights and revelations about who we are, and these insights often change as we grow and evolve. Our self-perception may shift as we encounter new experiences, meet different people, or face challenges that require us to question our assumptions about the world. Embracing this process with openness and curiosity can lead to profound personal growth and greater emotional resilience.

The Process of Self-Discovery

The path to self-discovery often begins with introspection—taking the time to reflect on one's thoughts, emotions, and behaviors. This reflection can help individuals uncover the motivations behind their actions and decisions. By understanding why we react in certain ways or are drawn to particular paths, we can begin to identify patterns in our thinking that influence our sense of self.

One of the first steps in the self-discovery process is to assess one's values and beliefs. What are the principles that guide your actions? What do you hold dear, and why? Understanding these core values is essential because they act as a compass, helping individuals make decisions that align with their authentic selves.

The Role of Self-Reflection in Self-Discovery

Self-reflection is a critical component of self-discovery, as it allows individuals to examine their actions, thoughts, and emotions from an objective standpoint. Through introspection, we gain deeper insights into our behaviors and reactions. This awareness can be transformative, as it allows us to identify areas of growth and potential for change.

Example: Emma's Journey of Self-Discovery

Emma, a successful professional in her early thirties, had always defined her worth by her career achievements. For years, she worked tirelessly to climb the corporate ladder, believing that success and validation would come from external recognition. However, Emma began to feel unfulfilled despite her accomplishments. She noticed that her personal life had been neglected, and she often felt disconnected from her true self.

After a period of self-reflection, Emma realized that her career, though important, was not her sole source of purpose. She identified her true passion for helping others and her desire to connect more deeply with people in meaningful ways. Emma's journey of self-discovery led her to pursue a career shift, leaving her corporate job to become a life coach. This new path allowed her to align her work with her values of connection and service, leading to greater fulfillment and authenticity in both her personal and professional life.

Emma's story exemplifies the transformative power of self-reflection. By taking time to pause and reassess her priorities, Emma was able to make significant changes that allowed her to live more authentically and in alignment with her true self.

Embracing Vulnerability in Self-Discovery

An often overlooked aspect of self-discovery is the willingness to embrace vulnerability. Vulnerability involves being open to experiencing and expressing emotions, particularly those that might feel uncomfortable or unfamiliar. It requires courage to expose our true selves, free from the facades we may put up to protect ourselves from judgment or rejection.

In the context of self-discovery, vulnerability means allowing oneself to be authentic, even in moments of uncertainty or discomfort. It involves accepting imperfections and being open to growth. By embracing vulnerability, individuals can cultivate deeper self-compassion and create a space where they can explore all aspects of themselves without fear of criticism.

Practicing Vulnerability

Identify a situation in which you typically guard yourself or avoid showing your true emotions. This could be in personal relationships, at work, or within your family.

In a safe environment, allow yourself to express your true feelings openly. Share something personal or let yourself be seen for who you truly are. Reflect on the experience and how it felt to let go of any protective barriers.

Embracing vulnerability is not about exposing yourself to harm or rejection, but about being honest with yourself and others. It is about allowing your true self to be seen and accepted, both by yourself and by those around you.

Self-Discovery through Challenges

One of the most powerful catalysts for self-discovery is facing challenges or difficult situations. When life presents obstacles, it often forces individuals to examine their beliefs, motivations, and sense of purpose. Challenges can expose hidden strengths, weaknesses, and desires that might otherwise remain dormant.

Through challenges, we may also discover new aspects of our identity. For example, overcoming personal adversity can reveal resilience, empathy, and a deeper understanding of our capabilities. While difficult experiences can be painful, they often offer some of the greatest opportunities for personal growth.

Michael's Transformation through Adversity

Michael's life was turned upside down when he lost his job in his late forties. For years, he had been comfortable in his career, but the sudden change forced him to confront questions he had avoided for years: What did he truly want out of life? What was his purpose beyond his job? After months of soul-searching and engaging in self-reflection, Michael discovered a passion for teaching and mentoring. He transitioned into a career as a life coach, helping others navigate life transitions. Michael's experience illustrates how challenges can be pivotal moments of self-discovery, enabling individuals to uncover new passions and redefine their sense of purpose.

The Lifelong Nature of Self-Discovery

Self-discovery is not a destination but a lifelong process. As individuals continue to grow and experience life, their understanding of themselves evolves. Each new stage of life presents fresh opportunities for self-exploration. The key is to remain open to these opportunities and continue the journey of introspection and growth.

By embracing self-discovery, individuals empower themselves to live authentically and align their actions with their deepest values. This journey of self-awareness not only enriches personal growth but also enhances relationships, enhances emotional well-being, and fosters a greater sense of purpose and fulfillment.

In conclusion, self-discovery is a profound and transformative journey that requires introspection, vulnerability, and openness to change. By reflecting on one's values, beliefs, and aspirations, individuals can uncover their true self and make decisions that align with their authentic desires. The process of self-discovery may be challenging at times, but it is ultimately a rewarding and empowering experience. As individuals continue to explore and embrace their true selves, they can live lives that are rich in meaning, purpose, and fulfillment.

Chapter 17: Celebrating Progress and Success

Celebrating progress and success is an essential aspect of the journey toward personal growth and transformation. While the focus is often placed on achieving goals, the process of acknowledging and celebrating the milestones along the way is just as important. Celebrating progress encourages a sense of fulfillment, enhances motivation, and reinforces the positive changes individuals make in their lives. As Maya Angelou wisely stated, "Nothing succeeds like success," emphasizing the power of recognizing and appreciating accomplishments, no matter how small. In this chapter, we explore the significance of celebrating progress and success, offering practical strategies and real-life examples to inspire readers to honor their achievements and embrace the journey of growth.

The Importance of Celebrating Success

Celebrating success isn't just about acknowledging the end result of hard work; it's about recognizing the effort, dedication, and resilience it took to get there. The act of celebration fosters a positive mindset, reinforces the idea that growth is a continuous process, and encourages individuals to keep pushing forward.

Often, people are so focused on the next goal or the future that they forget to appreciate how far they've come. Celebrating small wins and milestones provides a much-needed reminder of progress. This practice boosts confidence and provides the emotional energy to tackle new challenges. Moreover, celebrating achievements helps to build self-awareness, as it encourages individuals to reflect on their journey and acknowledge the growth they have experienced.

When individuals take time to appreciate their accomplishments, they affirm their capabilities, reinforcing the belief that they are on the right path. This sense of achievement can propel them to continue striving toward their goals and pursuing new aspirations.

Celebration as a Tool for Motivation

Celebrating success plays a pivotal role in maintaining motivation. By recognizing achievements, individuals can keep their energy and enthusiasm high. Celebration serves as a reminder of the progress made, which can be especially helpful during moments of doubt or frustration.

Practical Strategy: Creating Rituals for Success

- Create personal rituals to mark the completion of goalsor milestones. This could be as simple as enjoying a favorite treat after finishing a project, taking a moment to reflect on progress, or organizing a small gathering with friends or family.

- These rituals help mark the importance of the achievement and create positive associations with success. By making celebration a regular part of your routine, you cultivate an ongoing sense of motivation and fulfillment.

Additionally, taking time to reflect on progress allows individuals to recognize areas of strength and areas for improvement. This reflection is key for continued growth, as it helps inform future actions and decisions.

LAHARI KORUTLA

Sarah's Journey to Embrace Celebration

Sarah, a young entrepreneur, spent years building her small business. Despite the challenges she faced, she was often so focused on expanding her company that she rarely took the time to celebrate her wins. She felt that celebrating success would be premature until she reached her "big" goal—turning her business into a national brand.

One day, after securing her first major partnership, Sarah's mentor encouraged her to take a step back and reflect on her accomplishments. "Celebrating your progress isn't about waiting for the finish line," her mentor told her. "It's about acknowledging each step forward."

Taking this advice to heart, Sarah created a celebration ritual for every milestone. She began to celebrate when she hit a certain number of customers, completed a new product launch, or even improved her website's traffic. These moments of celebration boosted her confidence and kept her motivated. Sarah's story exemplifies the power of celebrating progress—by embracing the smaller victories, she was able to sustain her enthusiasm and grow her business with greater momentum.

The Role of Gratitude in Celebrating Success

Gratitude is another key element in celebrating progress. It shifts the focus from what is still lacking to what has already been achieved. Cultivating gratitude for one's journey not only enhances the celebration process but also fosters a sense of contentment and appreciation for life's experiences.

Practical Strategy: Integrating Gratitude into Celebration

- When celebrating a success, take a moment to reflect onthe steps that led to it. Consider the people who supported you, the lessons you learned, and the efforts you invested. This practice of gratitude not only enhances the celebration but also deepens the appreciation for the journey itself.

- You can also keep a gratitude journal, where you regularlynote down things you're grateful for. This could include both big achievements and small wins. By recognizing the contributions of others and appreciating the process, you can enrich the experience of success.

Gratitude can transform the celebration of success into a more meaningful and enriching experience. It reminds individuals of the value of their hard work, the lessons learned, and the relationships that supported their journey. By acknowledging both the small and significant moments, individuals are better able to embrace the fullness of their accomplishments.

Celebrating Progress and Personal Growth

Celebrating progress is not limited to the completion of specific goals. It also applies to the journey of personal growth. As individuals develop new skills, adopt healthier habits, or gain clarity in their lives, these moments of growth are worth celebrating as well. Recognizing these milestones reinforces the idea that growth is an ongoing process.

For example, if someone has been working on improving their mental health by practicing mindfulness or therapy, celebrating their progress might include acknowledging how far they have come in managing stress or improving emotional regulation. It's not just about achieving external markers of success but also about honoring internal transformations.

LAHARI KORUTLA

Celebrating these small, personal victories builds a strong foundation for continued growth. When individuals focus on their progress rather than perfection, they create a more positive and supportive environment for themselves.

John's Commitment to Personal Growth

John, a middle-aged man who had struggled with weight management for years, had always set lofty goals but never felt motivated to sustain his efforts. He would lose weight only to regain it, and this cycle left him feeling defeated. One day, John decided to shift his perspective and focus on celebrating his progress rather than fixating on the ultimate goal of reaching his ideal weight.

Each time he made a healthy choice, whether it was going for a walk, preparing a nutritious meal, or sticking to his exercise routine, John acknowledged it as a win. He celebrated these small victories by treating himself to a relaxing evening, sharing his progress with friends, or simply giving himself a pat on the back. Over time, this practice of celebration created a more positive mindset, and John's consistency improved. He didn't just focus on losing weight but on living a healthier, more balanced life, and his success became a reflection of his commitment to progress.

The Long-Term Impact of Celebrating Success

When individuals regularly celebrate their achievements, they create a feedback loop that reinforces positive behavior. Recognizing success helps build confidence, which, in turn, motivates individuals to pursue new goals and challenges. Over time, this cycle of progress and celebration fosters an enduring sense of accomplishment and fulfillment.

Celebrating success also creates a supportive environment for future endeavors. When individuals feel valued and appreciated for their hard work, they are more likely to maintain the momentum needed to continue their journey of personal and professional growth.

Honoring the Journey

In conclusion, celebrating progress and success is an essential part of personal transformation. By taking time to reflect on and appreciate achievements, individuals can boost motivation, reinforce positive habits, and deepen their sense of fulfillment. Gratitude plays a crucial role in this process, enhancing the celebration and fostering a deeper appreciation for the journey itself. Through real-life examples and practical strategies, it is clear that acknowledging success, no matter how small, empowers individuals to continue their path of growth with confidence and joy.

Conclusion:

As we reach the final pages of this book, we have embarked on a profound journey of self-discovery and growth. The lessons shared throughout have provided tools and insights to help you unlock the power of your mind and create a life that aligns with your true self. The mind is a dynamic and influential force, shaping the way we perceive and interact with the world. It is through intentional thought, self-awareness, and growth that we can navigate the challenges of life and cultivate the future we desire.

The process of personal transformation begins with an understanding that the mind is not fixed—it is malleable and adaptable. Each thought, belief, and action contributes to the creation of our reality. This book has presented strategies and techniques designed to help you harness this power. Whether through positive self-talk, goal-setting, or self-reflection, the tools are designed to empower you to shape your mindset and behaviors in ways that promote fulfillment and success.

One key concept explored in the book is the power of self-talk. Your internal dialogue has a profound impact on how you perceive yourself and the world. Positive self-talk not only boosts confidence but also influences how you approach challenges. It is essential to become aware of your thoughts and make conscious efforts to shift negative patterns that hold you back.

Consider the example of Sarah, a young professional who struggled with self-doubt. She was constantly questioning her abilities and fearing failure. Over time, she became aware of these patterns and decided to make a change. Sarah began practicing positive self-talk, replacing thoughts of inadequacy with affirmations of her worth and abilities. This shift in her inner dialogue led to greater self-assurance and a newfound motivation to pursue her goals. As she continued to reinforce this positive mindset, Sarah began to see real changes in her life, from her career advancement to her personal relationships.

This example shows the transformative impact that self-talk can have. It is a tool that, when used consciously, can unlock your full potential and create the foundation for lasting change. By cultivating a habit of positive thinking, you can reshape your experiences and build a life of resilience and fulfillment.

The pursuit of purpose is another critical aspect of personal growth. Life is most meaningful when it is guided by purpose, and when you are aligned with your core values. Through self-reflection, you can identify your passions and aspirations, and begin to make choices that align with your deeper sense of meaning. By setting purpose-driven goals, you can create a roadmap that brings you closer to living a fulfilling life.

Mark's story provides an example of the transformative power of purpose. Mark spent years in a job that provided financial stability but left him feeling unfulfilled. After some introspection, he realized that his true passion was education. He decided to transition into a teaching career, a choice that allowed him to align his work with his values. Mark's decision brought a new sense of energy to his life. His work became more meaningful, and he found greater satisfaction in helping others grow. This shift in his career also led to positive changes in other areas of his life, strengthening his relationships and improving his overall well-being.

Mark's story highlights the importance of understanding what drives you and making decisions that reflect those values. When you are aligned with your purpose, every action feels more intentional, and challenges become easier to navigate because they are connected to a greater sense of meaning.

Equally important is the practice of self-reflection, which allows for greater self-awareness and insight. By taking time to reflect on your experiences, behaviors, and emotions, you can gain a deeper understanding of what influences your decisions and reactions. Self-reflection enables you to recognize patterns in your thoughts and behaviors, and it gives you the clarity needed to make changes where necessary.

For example, imagine someone who has always struggled with maintaining close relationships. After engaging in self-reflection, they recognize that they often avoid vulnerability due to past hurts. This self-awareness allows them to make a conscious decision to practice openness and communication in their relationships. Over time, this shift leads to deeper connections with others and a more fulfilling social life.

This example demonstrates that through self-reflection, you can uncover the unconscious patterns that shape your life and take steps to alter them. Personal growth is a continuous journey of self-discovery, and the insights gained through reflection serve as a foundation for lasting transformation.

The role of habits in personal growth cannot be overstated. Transformation is often the result of small, consistent actions that compound over time. By developing habits that align with your values and goals, you can create lasting change in your life. Whether it's cultivating healthy routines, improving time management, or practicing gratitude, small actions performed consistently can lead to significant results.

Consider Anna, who sought to improve her physical health. For years, she struggled to maintain a fitness routine but always found excuses to put it off. One day, she made a commitment to start small—beginning with a 10-minute walk each morning. This simple habit quickly grew into a consistent exercise routine that she maintained over time. As Anna's fitness improved, she found that her energy levels increased, and she felt more confident in her body. By focusing on small, achievable actions, Anna was able to build a lasting habit that transformed her health.

Anna's story is a powerful reminder that transformation doesn't require drastic changes but rather the consistent application of positive habits. When you focus on building small habits that align with your goals, you set the stage for long-term success.

The final aspect of personal growth is celebration—acknowledging the progress you've made and honoring the milestones you've reached. Celebrating success, no matter how small, reinforces positive behavior and motivates you to continue moving forward. It also serves as a reminder of how far you've come and helps to cultivate a sense of gratitude for the journey.

Lisa, for instance, kept a journal to track her progress as she worked on her personal development. She would celebrate each small victory, whether it was standing up for herself in a meeting or practicing self-compassion during difficult moments. This practice of celebration not only kept her motivated but also helped her recognize the growth she had made. As she continued to celebrate her progress, Lisa found that her confidence grew, and she was more motivated to pursue new challenges.

Celebrating your achievements is not just about marking big victories—it's about acknowledging the journey and the steps that lead to success. By practicing gratitude and celebrating each milestone, you reinforce the positive habits that drive your growth.

LAHARI KORUTLA

Final Thoughts: Creating the Masterpiece of Your Life

In conclusion, the art of personal transformation is about understanding the power of your mind and the way it shapes your reality. By embracing tools such as positive self-talk, habit formation, self-reflection, and purpose-driven living, you begin to craft the masterpiece of your life. As with any piece of art, it requires time, patience, and dedication, but the results are worth the effort.

The mind is a canvas, and you are the artist. With each thought, action, and reflection, you add strokes of color to the canvas of your life. Whether through overcoming challenges, discovering your purpose, or embracing the power of habits, the path you choose is entirely yours to create. By aligning your actions with your core values and cultivating a mindset of resilience and growth, you can create a life that reflects the fullest expression of who you are meant to be.

The art of the mind is ever-evolving, and the journey of self-discovery and growth is never truly complete. Embrace each chapter, celebrate each step, and remember that the masterpiece you create is yours to shape. The journey may be long, but it is one of profound beauty, filled with endless opportunities for growth, fulfillment, and transformation.